NT

ELECTRONICS
SERVICING VOL. 2

ELECTRONICS SERVICING VOL. 2 Core Studies: Principles and Electronic Devices

K. J. Bohlman
I.Eng., F.S.E.R.T., A.M.Inst.E.

Dickson Price Publishers Ltd
Hawthorn House
Bowdell Lane
Brookland
Romney Marsh
Kent TN29 9RW

Dickson Price Publishers Ltd
Hawthorn House
Bowdell Lane
Brookland
Romney Marsh
Kent TN29 9RW

First edition published 1980
Second edition published 1989

British Library Cataloguing in Publication Data
Bohlman, K. J. (Kenneth John)
Electronics servicing. – 2nd ed.
Vol. 2: Core studies: Principles and electronic devices
1. Electronic equipment
I. Title II. Patchett, G. N. (Gerald Norman).
Electronics servicing 224 course for radio, television and electronics mechanics
621.381

ISBN 0-85380-191-6

Photoset by
R. H. Services, Welwyn, Hertfordshire
Printed and bound in Great Britain by
Biddles Limited, Guildford and King's Lynn.

CONTENTS

Other Books of Interest

ELECTRONICS SERVICING VOL 1
ELECTRONICS SERVICING VOL 3
ELECTRONICS SERVICING 500 MULTIPLE CHOICE QUESTIONS AND ANSWERS FOR PART 1
ELECTRONICS SERVICING 500 QUESTIONS AND ANSWERS FOR PART 2
COLOUR AND MONO TELEVISION VOL 1
COLOUR AND MONO TELEVISION VOL 2
COLOUR AND MONO TELEVISION VOL 3
PRINCIPLES OF DOMESTIC VIDEO RECORDING AND PLAYBACK SYSTEMS
VIDEO RECORDING AND PLAYBACK SYSTEMS 500 Q & A
CLOSED CIRCUIT TELEVISION VOL 1
CLOSED CIRCUIT TELEVISION VOL 2
RADIO SERVICING VOL 1
RADIO SERVICING VOL 2
RADIO SERVICING VOL 3
DIGITAL TECHNIQUES

Inspection Copies

Lecturers wishing to examine any of these books should write to the publishers requesting an inspection copy.

Complete Catalogue available on request.

AUTHOR'S NOTE

This volume, together with Vol. 3 (Principles and Electronic Circuits), will fully cover the Core Studies syllabus for Part 2 of the City and Guilds Electronic Servicing 224 Course.

CHAPTER ONE

BASIC THEORY OF SEMICONDUCTORS

MOST MATERIALS USED in electronic engineering fall clearly into the class of conductor or insulator. There are a number of materials in between these classes called semiconductors which have become vitally important and form the basis of the whole semiconductor industry.

PROPERTIES OF SEMICONDUCTOR MATERIALS

Resistivity

The name 'semiconductor' suggests that it has a conductivity lying somewhere between that of an insulator and that of a conductor. A good insulator such as polystyrene has a resistivity of the order of 10^{16}ohm-metre whereas the resistivity of a good conductor, e.g. copper, is approximately 10^{-8}ohm-metre. Semiconductor materials have intermediate resistivity values as shown in Fig. 1.1. For example, the semiconductor silicon of the type used in the manufacture of transistors may have a resistivity of about 1 ohm-metre at 20°C. Thus one property of a semiconductor lies in its resistivity value, typically in the range of 0·01 ohm-metre to 1 ohm-metre.

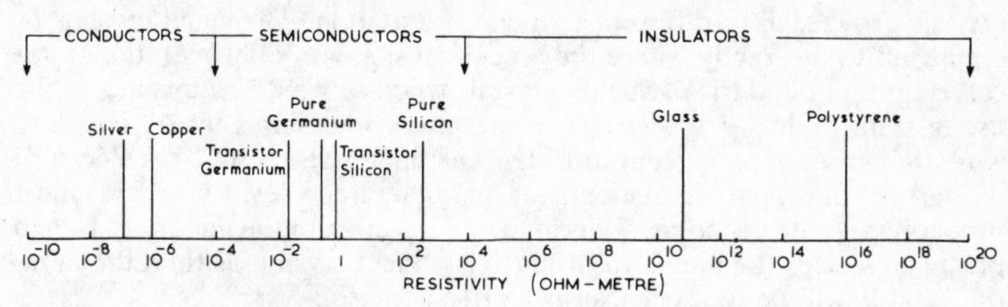

Fig. 1.1 Comparison of resistivity values.

Effect of Temperature

It would be wrong to assume that any material with a resistivity of this order is a semiconductor (this value could be fabricated by mixing materials

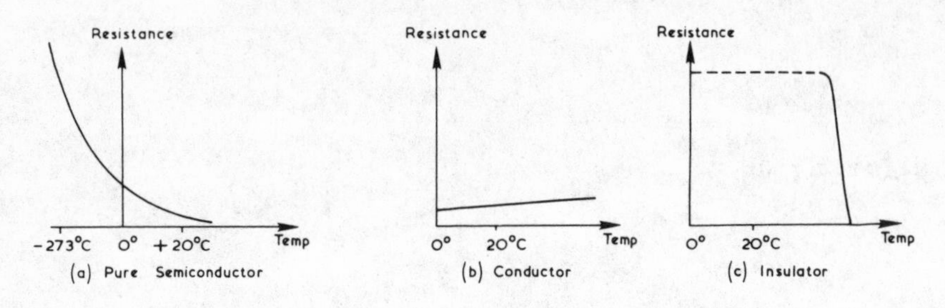

Fig. 1.2 Effect of temperature on resistance.

with high and low resistivities in suitable proportions). The semiconductor has other characteristics one of which is that it has a negative temperature coefficient of resistance, i.e. its resistance decreases with an increase in temperature. This is shown in Fig. 1.2(a), the resistance decreasing in a non-linear (exponential) manner as the temperature is raised. In contrast the resistance of a metallic conductor increases in direct proportion to the temperature, Fig. 1.2(b). This positive temperature coefficient of resistance is typical of all good conductors. The high insulation resistance of a good insulator does not appreciably change with temperature until a comparatively high temperature is reached at which point the resistance falls rapidly and the material becomes conductive (i.e. the insulator 'breaks down'). Like the semiconductor, the insulator exhibits a negative temperature coefficient of resistance, see Fig. 1.2(c). The difference between the insulator and semiconductor is therefore one of degree: the semiconductor being conductive at normal room temperature and the insulator at some higher temperature.

Crystal Form

The crystal structure of the materials used in the manufacture of semiconductor devices is rather important. When industrial metals such as copper solidify they produce a polycrystalline structure, i.e. many small crystals grow each of different size and orientation. The small crystals or grains fail to fit exactly where they meet causing areas between the grains called 'grain boundaries'. Such a crystal structure would be unsuitable for use in semiconductor devices since the grain boundaries would interfere with the motion of current through the material. Also, the electrical properties of apparently identical manufactured devices would most probably be quite different. Thus in the preparation of the materials used in practical devices the aim is to produce a perfect crystal, as then the more predictable will be its electrical properties.

The atoms in a perfect crystal arrange themselves into a regular geometric pattern which is repeated throughout the dimensions of the crystal. When the semiconductors germanium and silicon crystallise, the atoms arrange themselves into a regular cubic pattern or lattice as in Fig. 1.3.

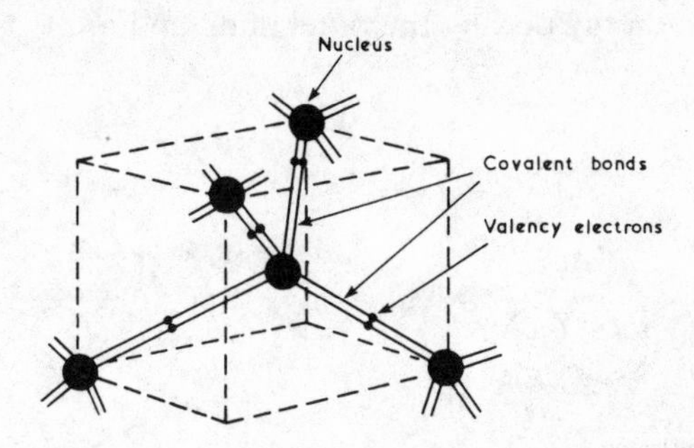

Fig. 1.3 Cubic crystal structure of germanium and silicon.

ATOMS

All materials are built up from molecules. A molecule is the smallest particle of material that can exist and still retain its original properties. In general, each molecule of a material is composed of smaller particles known as **atoms**. Materials formed by one type of atom only are called **elements**; there are approximately 100 elements known to science today. Some elements such as oxygen, hydrogen and carbon exist in large quantities and others are relatively rare, e.g. gold, plutonium and radium. In nature, one element rarely occurs separately from other elements. More often, elements exist in mixtures and in chemical combinations to form compounds. Water is a compound comprising two atoms of hydrogen and one atom of oxygen for each molecule.

Although different, atoms of all types of element have a common characteristic. They all consist of a **positively-charged central nucleus** around which rotate at a very high velocity one or more minute **negatively-charged** particles called **electrons**. The central nucleus contains two types of particles: **protons** exhibiting positive charges; and **neutrons** (except for the hydrogen atom) which are of zero electrical charge. The rotating electrons may all be in one orbit or a number of orbits. The number of protons and electrons in a normal atom are equal to one another and since their charges are equal in magnitude but opposite in sign they cancel out. Thus a normal atom is electrically neutral.

Examples of the structures of hydrogen and carbon atoms are shown in Fig. 1.4. The electrostatic attraction of the unlike charges of the electrons and protons provides the force required to maintain the electrons in their orbits. This force prevents the electrons from escaping from the atom, whilst the electron motion prevents them from plunging into the nucleus. With more complex atoms other orbits or 'shells' are used by the rotating electrons. The **maximum** number of electrons that can exist in any shell is the same for all elements, see Table 1. The first (inner) shell is full with 2 electrons, the second with 8 electrons, the third with 18 electrons and the fourth with 8 electrons. Electrons having the least energy occupy the inner shell and are tightly bound to the atom. On the other hand, electrons with the greatest energy occupy the outer shell and are loosely attached to the atom.

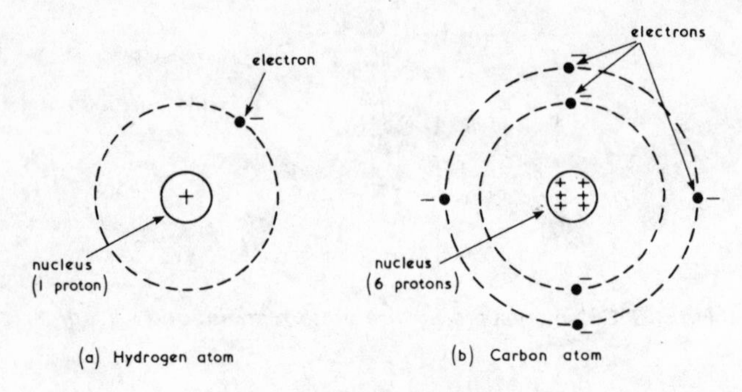

Fig. 1.4 Simple atomic structure.

TABLE 1

Shell	*Maximum number electrons for all atoms*	*Germanium atom (electrons per shell)*	*Silicon atom (electrons per shell)*
1st	2	2	2
2nd	8	8	8
3rd	18	18	4 (outer shell)
4th	8	4 (outer shell)	–

The electrons in the outer shell or orbit of an atom are called the 'valency electrons'. Since electrons in the outer shell are more easily detached from the atom it is the valency electrons that take part in chemical reaction and electrical conduction.

COVALENT BONDS

In the category of 'semiconductor' there are many materials. The most common devices – semiconductor diodes and transistors – are made from silicon and germanium and we will confine our attention to these particular materials.

In a germanium atom there are 32 electrons which are therefore grouped into four shells, see Table 1. The silicon atom has 14 electrons arranged in three shells. It will be noted from the table that germanium has an incomplete fourth shell (maximum number is 8 electrons) whilst silicon has an incomplete third shell (maximum number is 18 electrons) When a shell is incomplete it will try to make up the number from outside. One way of doing this is to form a 'covalent bond' with a neighbouring atom. The idea of covalent bonds is shown in Fig. 1.5 which is a two-dimensional representation of the crystal lattice structure of Fig. 1.3.

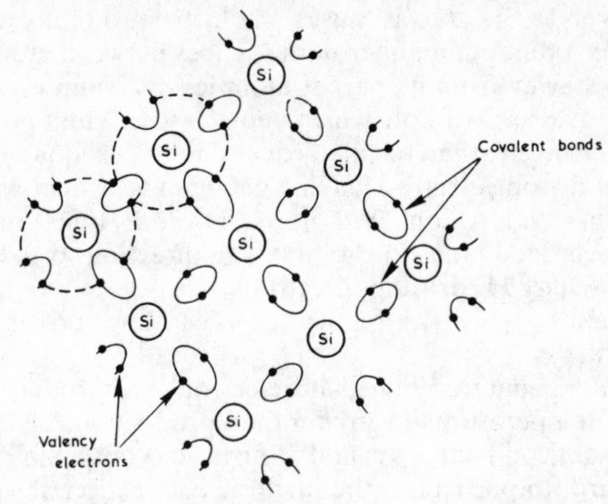

Fig. 1.5 Lattice of silicon atoms showing idea of covalent bonds (electron sharing) at −273°C.

The silicon atoms (Si) used in the diagram are shown only with their valency electrons; the inner shell electrons do not interest us. Each atom shares its valency electrons with neighbouring Si atoms thus in some respects the atoms behave as if they had 8 electrons in their outer shells. The third shell is stable when either 8 or 18 electrons are present. It is the sharing of valency electrons by neighbouring atoms that provides the covalent bonds within the crystal and keeps the atoms stable. In the diagram all the electrons are engaged in forming covalent bonds thus there are no current carriers available, i.e. the material is acting like an insulator. This condition is true only at absolute zero (−273°C).

CONDUCTION IN PURE SEMICONDUCTORS

When energy is supplied to the crystal either in the form of heat or light, the atoms vibrate and some of the covalent bonds are broken. Thus the condition illustrated in Fig. 1.5. cannot be realised in practice since heat is everywhere.

At normal room temperature (+20°C) the occasional bond is broken and the situation may be expressed diagrammatically as in Fig. 1.6. The thermal

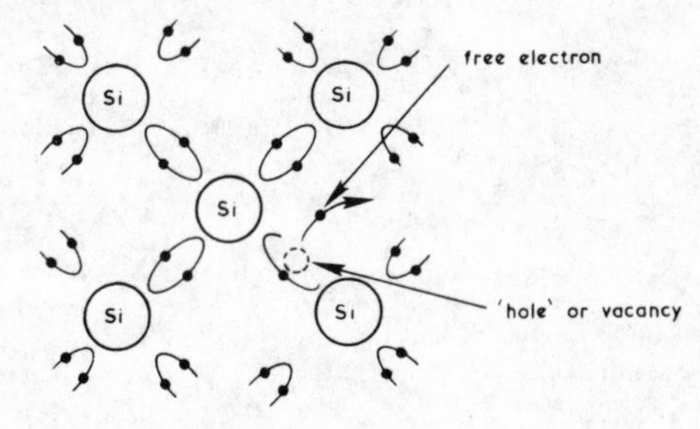

Fig. 1.6 Conduction in pure semiconductor (breaking of covalent bonds at 20°C).

energy given to the crystal causes an electron to break away from its orbit and drift in a random manner in the spaces between atoms. Every electron that breaks away from its parent atom leaves behind it a vacancy called a 'hole'. This vacancy or hole which behaves with a unit positive charge equal to one electron can then be filled (due to the attraction of opposite charges) by another drifting electron leaving behind it a vacancy which thus becomes filled again, and so on. We may therefore think of a movement of positively-charged holes in the opposite direction to the movement of the negatively-charged drifting electrons. Holes move more slowly than electrons and so under the application of a p.d. would carry less than 50% of the current.

Thermally generated holes and electrons are produced in pairs and the higher the temperature the greater their production. Thus the conduction in a pure semiconductor, called 'intrinsic conduction', increases with temperature. In particular, the intrinsic conduction of silicon is very much less than that of germanium at a given temperature. This is because more energy is required to break the covalent bonds of silicon atoms than germanium atoms.

CONDUCTION IN DOPED SEMICONDUCTORS

The materials in the state so far described are seldom used in practical devices since the conductivity at normal temperatures is not high, the conduction is very sensitive to temperature changes and is composed partly of electrons and partly of holes. By the controlled addition of minute traces of impurity to the intrinsic material these disadvantages may be overcome.

N-type Material

If an impurity with atomic dimensions similar to those of the intrinsic material are introduced into the pure material, the impurity atoms can take up a position without seriously disturbing the crystal lattice. The idea is shown in Fig. 1.7(a) where an impurity atom of phosphorous (P) is 'sitting' in

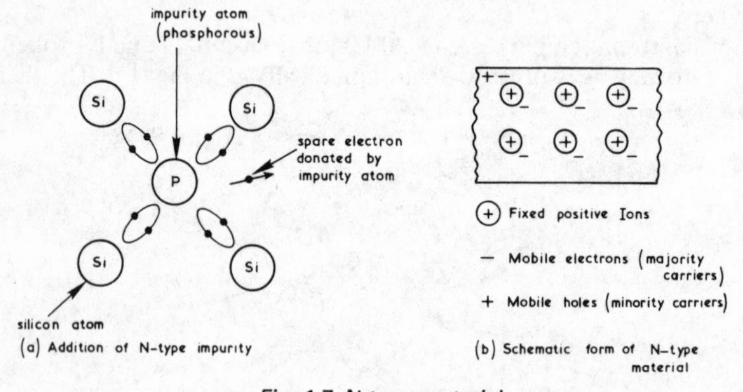

Fig. 1.7 N-type material.

the crystal lattice of silicon atoms. Now, phosphorous has five electrons in its outer shell, thus four of its valency electrons will readily form bonds with neighbouring silicon atoms but the fifth electron is left spare. It is found that very little energy is required to detach the spare electron from its parent atom. The thermal energy at room temperature is sufficient to part most of the spare electrons from their impurity atoms and the electrons can then be used as current carriers.

The amount of impurity added is of the order of 1 impurity atom to every 10^8 intrinsic atoms. Even this minute trace of impurity is sufficient to increase the conductivity by a factor of about 15. The process of adding controlled levels of impurity is called 'doping'. In this case the doping has bestowed N-type conductivity since the charge carriers are negatively-charged electrons. The impurity atoms are sometimes called 'donor atoms' as they give electrons to the lattice. Other examples of N-type impurities (those with five valency electrons) are arsenic and antimony.

When an impurity atom loses an electron its electrical neutrality is upset. Having lost an electron the impurity atom exerts a positive charge and becomes what is called a 'positive ion'. The ion does not move as it is held immobile by the covalent bonds; only the donated electrons are free to move. In diagrams, N-type material is often represented as shown in Fig. 1.7(b). It should be remembered that there will also be electrons and holes produced by the breaking of bonds of the intrinsic atoms as previously explained. However, in N-type material at normal temperatures the density of the donated electrons is greater than the thermally generated holes. Thus in N-type material the electrons are referred to as the **majority carriers** and the holes the **minority carriers**.

P-type Material

The opposite type of conductivity can be obtained by adding minute quantities of impurities that have only three electrons in their outer shells. The idea is shown in Fig. 1.8(a) using a boron (B) impurity atom. The three valency electrons of the boron atom will readily form covalent bonds with

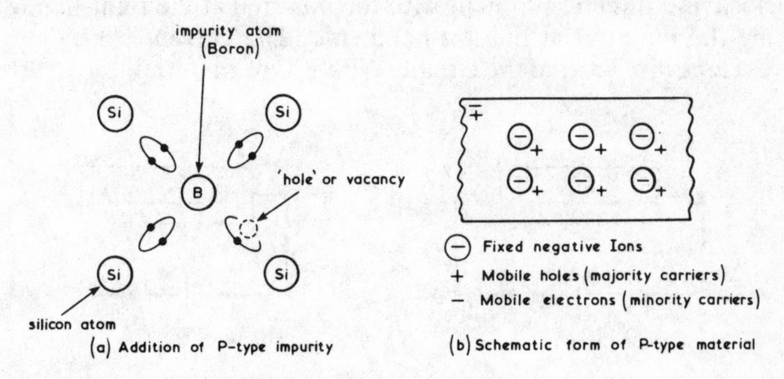

Fig. 1.8 P-type material.

the valency electrons of three neighbouring silicon atoms. However, the fourth lattice valency is short of an electron and this vacancy behaves like a hole, i.e. with a positive charge influence. Very little energy is required to attract a nearby valency electron (from a silicon atom) to the hole which thus becomes filled and the hole moves away from the impurity atom (to a silicon atom). At room temperature the thermal energy given to the crystal is sufficient to fill nearly all the holes created by the addition of the impurity.

In this case the doping has given P-type conductivity to the pure material since the carriers are positively-charged holes. The impurity atoms are called 'acceptors' because the holes created accept electrons from the crystal lattice. Other P-type impurities are gallium and indium.

When the impurity atom gains an electron its electrical neutrality is upset. Having gained an electron the impurity atom exerts a negative charge and becomes a 'negative ion'. Again, the ion does not move as it is held immobile by the covalent bonds; only the holes are free to move. In diagrams, P-type material is often represented as in Fig. 1.8(b). As for the N-type material there will also be electrons and holes produced by the breaking of bonds in the intrinsic material. However, in P-type material at normal temperatures the density of the holes is greater than that of the thermally generated electrons. Thus in P-type material the holes are referred to as the **majority carriers** and the electrons as the **minority carriers**.

The special conductivity of both N-type and P-type materials is called 'extrinsic conduction' since it is due to the addition of impurities to the pure semiconductor. In their separate forms both P-type and N-type materials are electrically neutral since the charge of the fixed ion is balanced by the opposite charge of the mobile carrier.

Current Flow

If a p.d. is applied to an extrinsic semiconductor there will be a drift of charge carriers from one end of the material to the other. With N-type material and an external battery arranged as in Fig. 1.9(a) there will be a drift of electrons from right to left within the material. In the wires of the external circuit, current is supported by the movement of electrons in an anticlockwise direction moving into the material at the right-hand side and leaving the material at the left-hand side. The current (I) by convention moves clockwise around the circuit. With P-type material, Fig. 1.9(b), holes

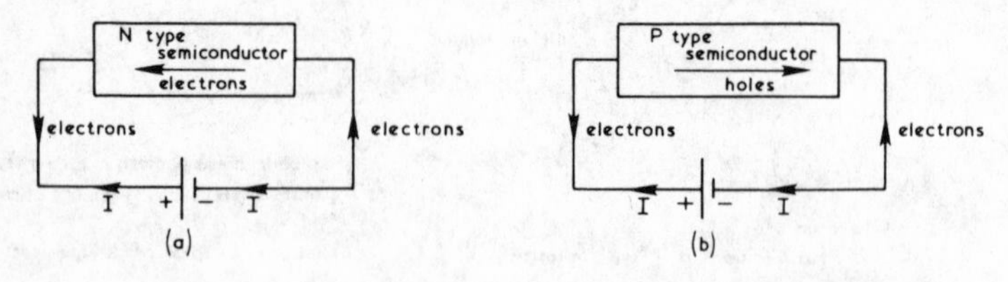

Fig. 1.9 Current flow in N-type and P-type semiconductors.

move from left to right within the material. In the external circuit however, current can only be supported by electron movement and for every hole arriving at the right-hand side of the material an electron is received from the external circuit. It will be noted that conventional current (I) is in the same direction as hole movement.

QUESTIONS ON CHAPTER ONE

(1) The resistivity of semiconductors is typically:
- (a) 10^{-6} to 10^{-8} ohm-metre
- (b) 10^{6} to 10^{8} ohm-metre
- (c) 10^{12} to 10^{14} ohm-metre
- (d) 0·01 to 1 ohm-metre.

(2) A positive temperature coefficient of resistance is typified by:
- (a) Insulators
- (b) Conductors
- (c) Semiconductors
- (d) Silicon.

(3) An atom consists of:
- (a) A negatively-charged nucleus and orbiting positive charges
- (b) A positively-charged nucleus and orbiting negative charges
- (c) A positively-charged nucleus and orbiting neutrons
- (d) Electrons and holes in equal numbers.

(4) The semiconductor silicon atom has:
- (a) 3 valency electrons
- (b) 4 valency electrons
- (c) 8 valency electrons
- (d) No valency electrons.

(5) Intrinsic conduction occurs only in:
- (a) Pure and doped semiconductors
- (b) Doped semiconductors
- (c) Pure semiconductors
- (d) Silicon.

(6) Extrinsic conduction occurs only in:
- (a) Pure and doped semiconductors
- (b) Doped semiconductors
- (c) Pure semiconductors
- (d) Silicon.

(7) The majority charge carriers in N-type material are:
- (a) Electrons
- (b) Holes
- (c) Positive ions
- (d) Negative ions.

(8) The addition of an impurity atom having five valency electrons to pure germanium will cause the material to become:
- (a) Non-conductive
- (b) P-type
- (c) N-type
- (d) More thermally sensitive.

(Answers on page 180)

CHAPTER TWO

SEMICONDUCTOR DIODES

A SEMICONDUCTOR OR p-n diode device is obtained from the junction formed by separate areas of extrinsic germanium or silicon one of which has P-type conductivity and the other N-type, Fig. 2.1. Nearly all the important effects that occur in semiconductor devices take place in the region of the

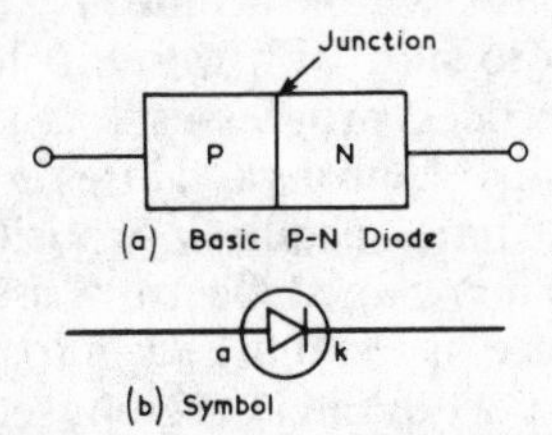

Fig. 2.1 Semiconductor diode.

junction between different materials. The most important of these is the low resistance offered by the junction to the flow of current in one direction and high resistance in the other direction, a characteristic that is always present.

THE P-N JUNCTION

There are a number of methods used in the fabrication of p-n junctions and some will be described later. Current techniques use a single crystal to form the junction with suitable 'doping' applied to provide a change in the semiconductor from N-type to P-type. This change may be abrupt as in Fig. 2.2 or may occupy many lattice spacings of the crystal. A junction cannot properly be formed by taking a piece of N-type material and joining it to a section of P-type.

Figure 2.2 shows a small section of a p-n diode in the vicinity of the junction but at this point no external connections are made. When the junction is first formed, larger numbers of electrons wander from left to right across the junction than in the opposite direction since the N-region contains many electrons but the P-region only a few. Similarly, more holes wander across the junction from right to left rather than in the opposite direction as the P-region contains many more holes than the N-region. This initial

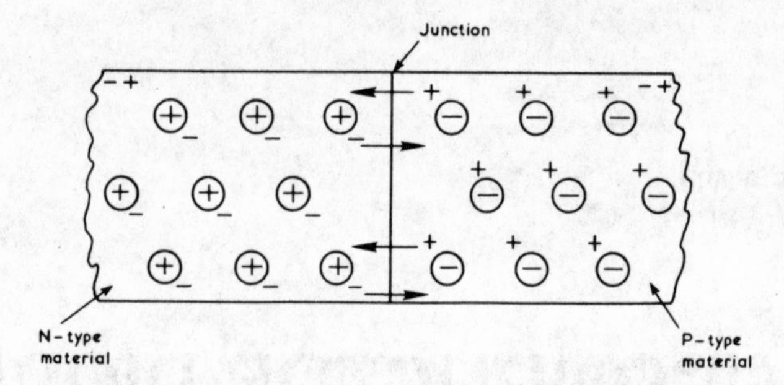

Fig. 2.2 Small section of p-n junction (showing initial diffusion of majority carriers).

exchange of carriers across the junction is due to 'diffusion', a process whereby electrons and holes spread out from their respective concentrations to try to occupy the space uniformly. Another example of diffusion is the mixing of a dye (e.g. ink) in an unstirred glass of water. This initial diffusion process in the p-n junction is not electronically based.

The two separate areas to start with were of the same potential (zero). However as the diffusion process progresses the N-side gains holes and loses electrons and the P-side gains electrons and loses holes. The N-region having lost electrons and gained holes acquires a positive charge whereas the P-region having lost holes and gained electrons assumes a negative charge. Thus a potential difference is established across the junction with the P-region negative with respect to the N-region, see Fig. 2.3(a). This p.d. is called the 'diffusion p.d.' or 'barrier p.d.' Majority electrons and holes from either side now crossing the junction have to work **against** a rising p.d. Some of these carriers will have greater energy than others, overcoming the p.d.

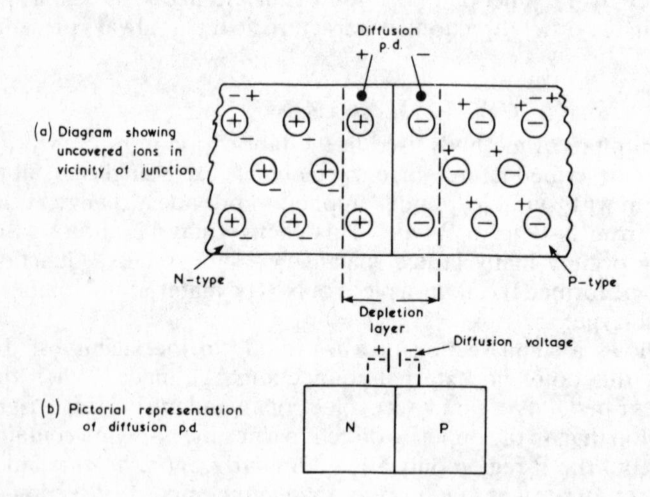

Fig. 2.3 Diffusion p.d. and depletion layer.

and so cross the junction thereby causing the p.d. to rise. This will reduce the numbers of majority carriers subsequently crossing and the action will become self-limiting. Pictorially, the diffusion p.d. may be represented by a fictitious battery as in Fig. 2.3(b). The diffusion p.d. cannot be measured with a voltmeter; when connections are made to the N and P sides two p.d.s will appear at the connections. These will have a net value that will exactly balance the diffusion p.d. This must be so otherwise the diffusion voltage would be a perpetual source of energy.

Depletion Layer

When the majority carriers diffuse across the junction they recombine thereby neutralising one another. This causes a lack of charge carriers in the immediate vicinity of the junction and the immobile ions are 'uncovered', see Fig. 2.3(a). This area of uncovered ions (positive ions on the N-side and negative ions on the P-side) is called the 'depletion layer' because it is depleted of mobile charge carriers. Typically the width of the depletion layer is about 0·0001 cm, or 1μm.

Junction Under Bias Conditions

Figure 2.4(a) represents a p-n diode in schematic form with **ohmic** connections made to the separate areas. An ohmic connection is a non-rectifying one.

Reverse Bias

When an external supply (V_r) is connected with polarity as indicated in Fig. 2.4(b), the junction is said to be 'reverse biased'. Under this condition the external voltage subjects the majority carriers of each region to an electric field which draws them away from the junction. This action causes more immobile ions to be uncovered thereby increasing the width of the depletion layer and so increasing the diffusion p.d. Due to the rise in the diffusion p.d., majority carriers (electrons from the N-region and holes from the P-region) find it more difficult to cross the junction and in consequence the actual numbers of majority carriers crossing decrease. The rise in diffusion p.d. will however aid the passing of minority carriers (holes from the N-region and electrons from the P-region). The net current crossing the junction is composed of a small number of minority carriers as indicated by the arrow heads in the diagram. This condition corresponds to the 'non-conducting' or 'cut-off' state of the p-n diode when it may be regarded as almost an open circuit. When the reverse voltage is large, the rise in diffusion p.d. reaches a point where no majority carriers cross the junction. In this case only minority carriers make the crossing, but they are small in numbers.

In the external connecting wires, current can only be supported by electron movement. Thus when minority electrons crossing from right to left

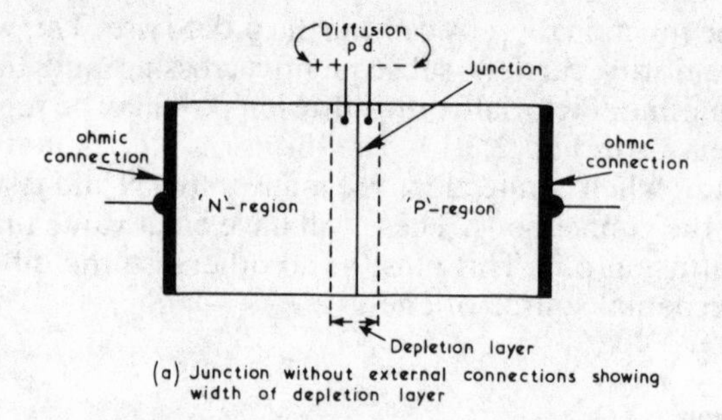

(a) Junction without external connections showing width of depletion layer

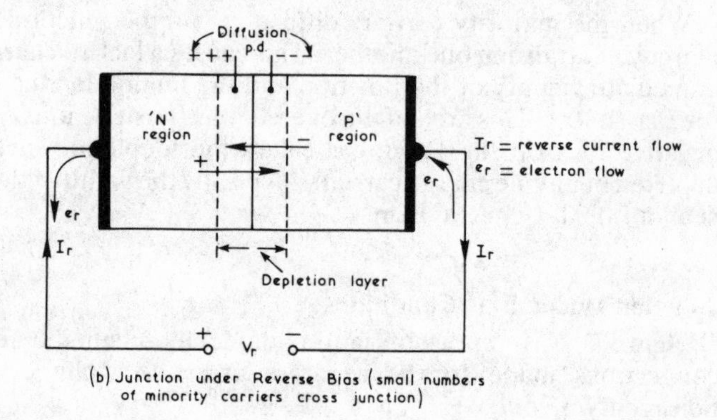

(b) Junction under Reverse Bias (small numbers of minority carriers cross junction)

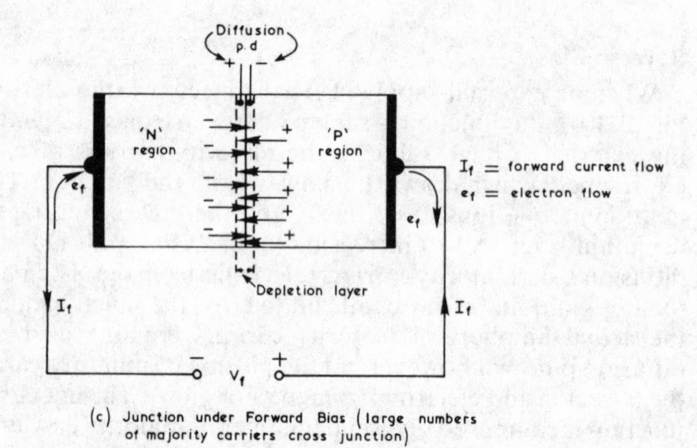

(c) Junction under Forward Bias (large numbers of majority carriers cross junction)

Fig. 2.4 Diode-like property of p-n junction.

arrive at the left-hand ohmic connection they will pass into the connecting wire and move towards the positive terminal of the supply. When minority holes moving from left to right reach the right-hand ohmic connection they will receive electrons from the negative terminal of the supply and become neutralised.

Forward Bias

If the polarity of the external voltage is changed to that shown in Fig. 2.4(c), the junction is said to be 'forward biased'. In this condition the external voltage (V_f) exerts a force on the majority carriers urging them closer to the junction thereby reducing the number of 'uncovered' ions. In consequence, the width of the depletion layer is reduced which causes the diffusion p.d. to fall. As a result large numbers of majority carriers are able to cross the junction since majority electrons and holes of lower energy will find the diffusion p.d. surmountable. A large current now flows across the junction and this condition corresponds to the 'conducting' or 'on' state when the junction is of low resistance.

Majority carriers crossing the junction are indicated by the arrow heads in Fig. 2.4(c). The majority electrons crossing from the left will recombine with holes in the P-material almost immediately they have crossed the junction. For every recombination that occurs an electron is given up from an area close to the right-hand ohmic connection to the positive terminal of the supply. For every electron 'sucked-off' in this way by the supply, a hole is created which moves towards the junction. Majority holes crossing the junction from the right will recombine with electrons in the N-material. For each recombination that takes place, an electron is given up by the negative terminal of the supply, enters the crystal by way of the left-hand ohmic connection and subsequently moves towards the junction. Thus again in the external circuit, current is supported by electron movement only. Note however that the large forward current (I_f) is in the opposite direction to the small reverse current (I_r).

The term 'diode' in p-n diode means two electrodes which are called 'anode' and 'cathode' (as used in the thermionic diode valve). To place a p-n diode in the forward bias condition or low resistance state, the anode terminal *a* of Fig. 2.1(b) must be made positive with respect to the cathode terminal *k*. The arrow in the symbol points the direction of conventional current flow.

Typical Characteristic of P-N Diode

The graph of Fig. 2.5 shows the relationship between the current in a p-n diode and the voltage drop across it. As the voltage is increased from zero in the forward direction, the diffusion p.d. of the junction is reduced and majority carriers cross the junction in increasing numbers. Further increases in forward voltage increases not only the number of carriers crossing but also their speed of travel. Thus, as the diffusion p.d. is overcome by the applied voltage and it falls almost to zero, the forward current rises rapidly in exponential form.

When the polarity of the applied voltage is reversed and gradually increased, the current rises a little at first and then settles down to a constant but very small value. At this steady level, the current passing across the junction is due entirely to minority carriers. The characteristic illustrates the almost perfect rectifier action of the p-n diode.

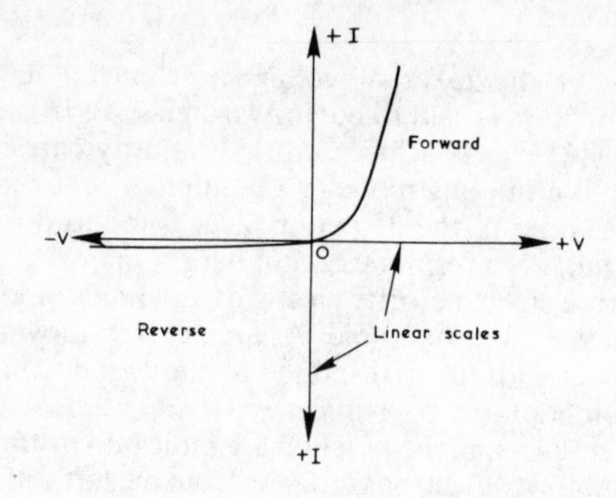

Fig. 2.5 Voltage–current characteristic of p-n diode.

For practical purposes the scales of Fig. 2.5 are adjusted so that we can see more clearly the usual high-current, low-voltage operation in the forward direction and the high-voltage, low-current operation in the reverse direction. In Fig. 2.6 the scales have been adjusted. The forward current commences its steep upward swing at a very low forward voltage of somewhere between 0·2V and 0·3V for a germanium device and 0·6V to 0·8V for a silicon diode. At these low forward voltages a significantly large current can be sustained. In the reverse bias state the reverse current saturates in this example at about 4μA. If this value is compared with a forward current of say 1A, the forward current is $2 \cdot 5 \times 10^5$ greater than the reverse flow. Here, we are using a Ge diode as an example; in a silicon type the reverse current is very much smaller.

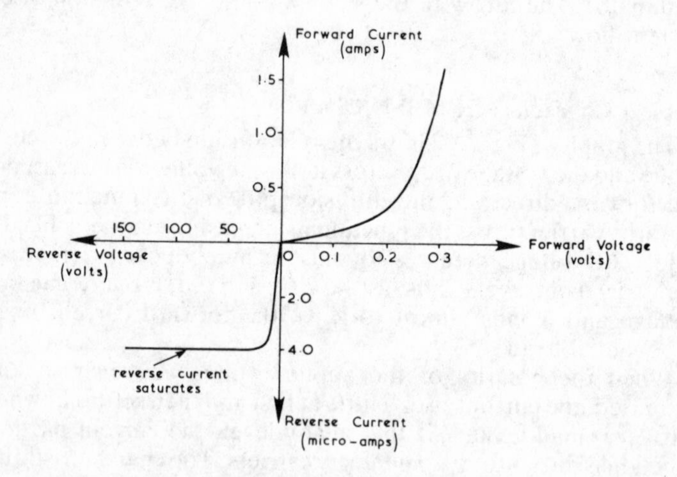

Fig. 2.6 Voltage and current scales adjusted.

Effect of Temperature

The effect of a rise in temperature on the characteristics of a p-n diode (Ge type assumed) is illustrated by Fig. 2.7. Raising the temperature increases the number of minority carriers generated and as these carriers are responsible for the reverse current the reverse saturation level will increase. Roughly, for each increase of 5°C for silicon and 8°C for germanium the reverse current is doubled. Although the rate of increase is greater for silicon, the absolute numbers of minority carriers are very much less than for germanium. Therefore, silicon is chosen for semiconductor junctions which have to work at high temperature (150°C max. for silicon and 85°C max. for germanium).

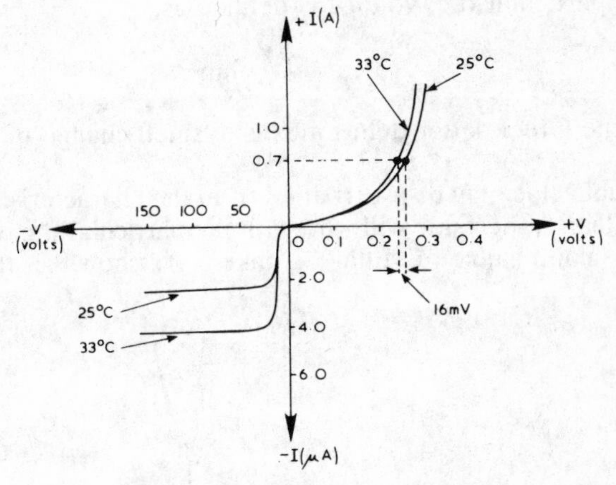

Fig. 2.7 Effect of temperature rise on characteristic.

In the forward direction, the thermal energy given to the device during a temperature rise increases the energy of the carriers and there is a rise in the forward current. When maintaining a given current the **voltage drop** across the diode decreases by approximately 2mV for every degree C rise in temperature for germanium and silicon. In the example, with a given current of 0·7A and a temperature rise of 8°C the voltage drop across the diode decreases by 16mV.

The temperature of a junction may rise due to an increase in the surrounding air temperature (ambient temperature) or as a direct result of power dissipated within the device under forward or reverse bias conditions. In both cases the rise of current that follows modifies the voltage–current characteristic. An increase in reverse current causes a rise in the power dissipated and hence an increase in junction temperature. This creates more minority carriers causing an increase in reverse current, the power dissipated rises producing a further increase in the junction temperature, and so on. The current continues to rise as the temperature of the junction grows, creating a cumulative effect known as 'thermal runaway', which may eventually destroy the diode unless the current can be limited. This is a condition most likely to arise under high ambient temperature or when the heat transfer (heat sink) is inadequate.

A.C. or Slope Resistance

The **d.c. resistance** of a diode is simply the ratio of the voltage drop (V) across the device to the current (I) passing through it and is measured in ohms. Since the characteristic of the p-n diode is not linear, the d.c. resistance will vary with the point of measurement. Generally, we are interested in the value of resistance offered by active devices such as the p-n diode to **changes in voltage**, i.e. **a.c. resistance**. The a.c. resistance (r_a) of a diode at a particular d.c. voltage is defined as

$$r_a = \frac{\delta V}{\delta I} \text{ ohm}$$

where δ (the Greek letter delta) means a 'small change of'.

The actual value may be ascertained from the characteristics as shown in Fig. 2.8. Clearly, the value will vary with the particular d.c. voltage around which the small change of voltage is taken. As shown, if the same 'small

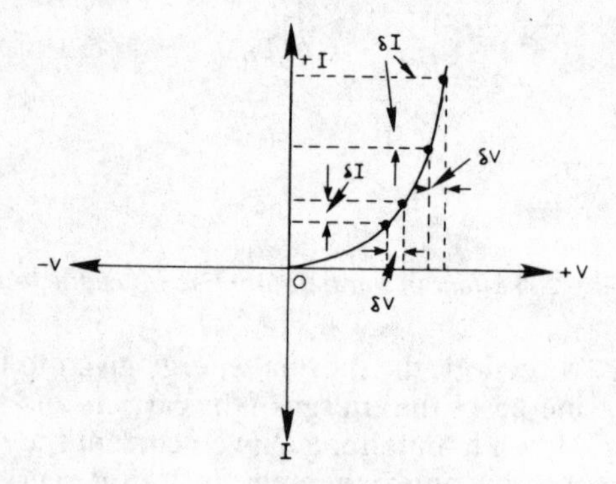

Fig. 2.8 A.C resistance of diode.

change of voltage' is taken further up the characteristic the resultant 'small change of current' will be greater. As a result the a.c. resistance is less when the slope of the characteristic is large. In the forward direction its value may vary from tens of ohms to a fraction of an ohm. In the reverse direction, where the slope of the characteristic is extremely small, the a.c. resistance is of the order of thousands of ohms.

Construction of Junction Diodes

There are two main techniques used one of which is shown in Fig 2.9. A small pellet of indium is placed on a thin wafer of N-type germanium and then heated to a temperature above the melting point of indium but below

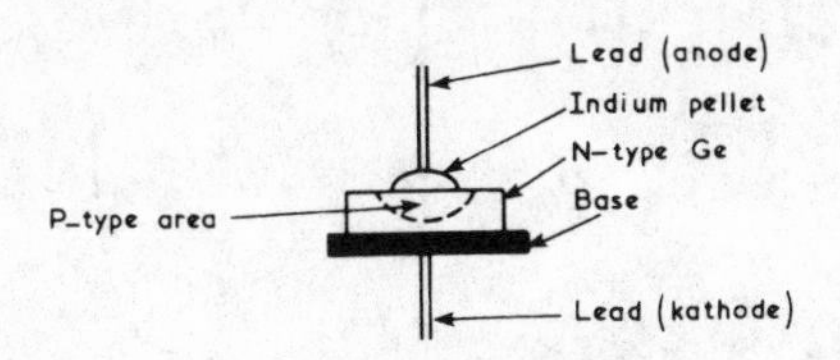

Fig. 2.9 Alloy junction diode.

that of germanium. This causes the indium (a P-type impurity) to alloy with the germanium producing on cooling an area in the wafer having P-type conductivity. Thus we now have a germanium wafer that is part P-type and part N-type. The wafer is soldered to a copper or brass base (to provide good heat transfer) using a solder rich in antimony. The antimony being an N-type impurity prevents the formation of another junction during the process of soldering, i.e. an ohmic contact is created. Leads are then attached to the indium pellet and base. The device is then hermetically sealed in a case of metal and glass or metal and ceramic to exclude moisture and impurities which can cause deterioration. This method is known as the 'alloying process'. Silicon alloy juction diodes can be made in the same way using a wafer of N-type silicon and a pellet of aluminium (a P-type impurity).

The other technique is known as the 'diffusion process' and the basic idea is shown in Fig. 2.10 for a germanium juction. P-type germanium is heated to

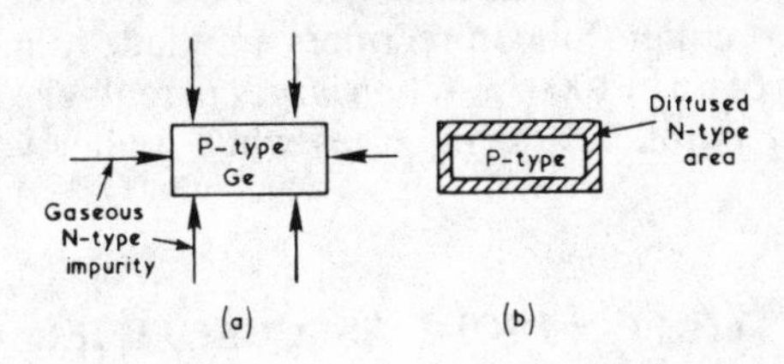

Fig. 2.10 The diffusion method of producing a p-n junction.

a high temperature (just below the melting point of germanium) and at the same time exposed to an N-type impurity in gaseous form (the element antimony may be used), see Fig. 2.10(a). The gaseous antimony diffuses into the germanium to produce an N-type area as shown in Fig. 2.10(b). If a silicon device is required, phosphorous in gaseous form is allowed to diffuse into a section of P-type silicon. The rate at which the impurity gas diffuses depends upon the temperature and may take up to 1–2 hours. This process is more easily controlled than the alloying method.

Silicon junction rectifiers are commonly manufactured using the diffusion process and they are used almost exclusively in the power section of radio and television receivers and other electronic equipment. A typical characteristic of a silicon junction rectifier is given in Fig. 2.11. It is similar to a germanium juction diode (see Fig 2.6) but there are important differences. The forward characteristic is more non-linear and commences its abrupt upward swing at about 0·6 to 0·8V. For a small diode passing about 10mA

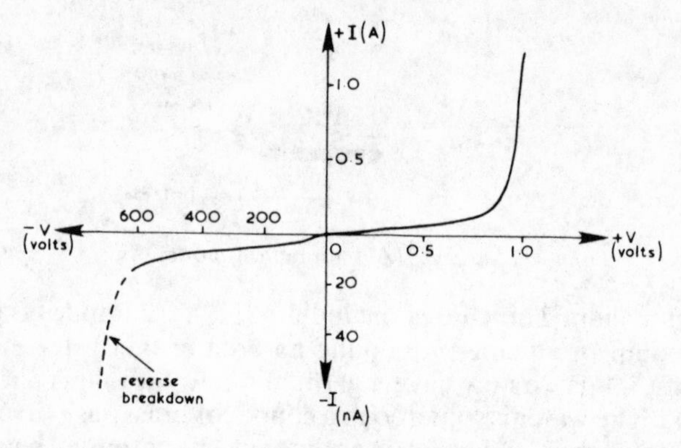

Fig. 2.11 Typical characteristic of silicon junction rectifier.

the forward voltage drop may be about 0·6V, whereas for a larger type passing say 1A the voltage drop may be about 1·0V. In the reverse direction the current is extremely small and for most practical purposes it is negligible; note that the reverse current scale is nA (10^{-9}A). With any junction diode, if the applied reverse voltage is too high is may cause the junction to break down causing a large reverse current to flow (shown dotted). There is, therefore, a maximum voltage that may be applied in the reverse direction. For a silicon rectifier this may be as high as 2000V as opposed to about 500V for a germanium rectifier. Silicon rectifiers are made to pass currents in the forward direction of up to 1000A. When large currents are required rectifiers may be placed in parallel, and in series when higher voltage ratings are required.

THE P-N DIODE AS A RECTIFIER

In Volume 1 of this series a rectifying unit was shown as a block forming part of a power supply. It is in power supplies that p-n diode(s) may be used as the rectifying unit to convert a.c. to d.c. A simple rectifier circuit (half-wave) is shown in Fig. 2.12 using a p-n diode and a load (represented by a resistor) fed with an alternating voltage at its input.

When terminal A is positive to terminal B during the first half-cycle of the input, the diode is placed in its 'forward' or low resistance state. Thus a current will flow from terminal A through the diode and load and back to the supply at B. During the following half-cycle when the input is negative, terminal A is negative to terminal B and the diode is placed in its 'reverse' or high resistance state. If we ignore the small reverse current that flows, no current flows during the negative half-cycle. On the second positive half-cycle of the input, the diode is again placed in the 'forward' state and current flows in the rectifier and load once more. During the second negative half-cycle the diode is placed in the 'reverse' state and again no current flows. Thus the current in the load and hence the voltage across it consist of half-cycles as shown. This is d.c. since the half-cycles are acting in one

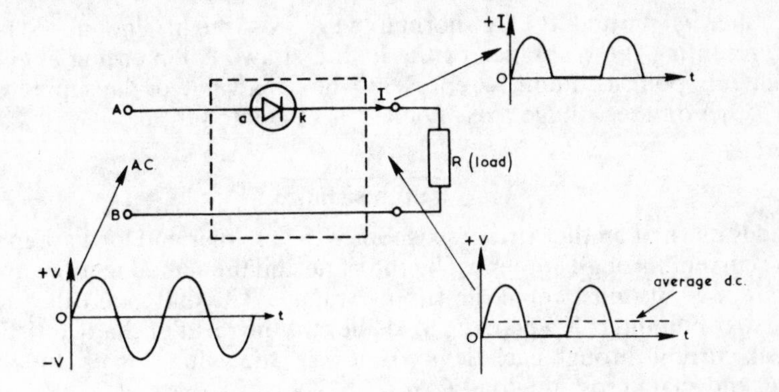

Fig. 2.12 Use of p-n diode in half-wave rectifier circuit.

direction only (positive); they would be in the opposite direction (negative) if the connections of the diode were reversed. The d.c. as it stands would not be smooth enough to provide a d.c. supply for electronic equipment so a smoothing circuit would be required. Further details of complete rectifier circuits including smoothing components are given in Volume 3 of this series.

Current in Rectifier and Resistive Load

The voltage drop across a semiconductor rectifier may be quite small compared with that across the load in some power supplies, in which case it may be neglected. However, it is useful to consider how the value of the peak current flowing in the rectifier and resistive load may be obtained when the voltage drop of the rectifier cannot be neglected.

One method is shown in Fig. 2.13 where the *V–I* characteristic of the

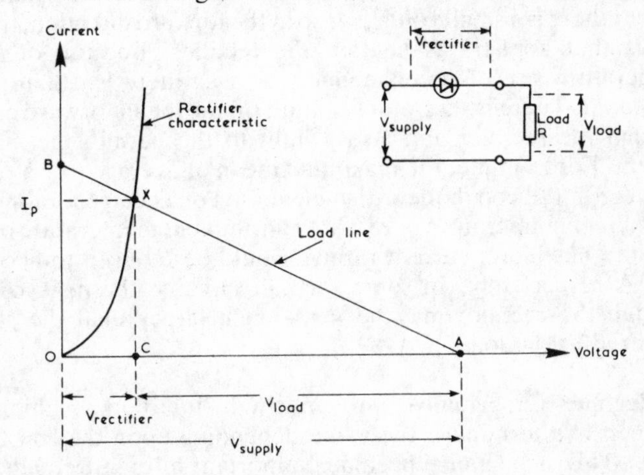

Fig. 2.13 One method of finding the peak current in a rectifier circuit.

rectifier is plotted in the normal way. A straight line or 'load line' representing the resistance of the load is drawn commencing at A and is joined to point B. Point A represents the peak value of the supply voltage (V_{supply}) on the voltage axis. Point B is calculated from

$$\frac{V_{supply}}{\text{load resistance}}$$

and is marked on the current axis. Since the rectifier and load are in series, the current through both must be the same and the sum of rectifier and load voltage drops must equal the supply voltage. The only point that satisfies these conditions is X where the load line cuts the rectifier characteristic. The peak current through each device is now I_p, the voltage across the rectifier OC and that across the load CA.

Choice of Rectifier

When choosing a semiconductor rectifier for a particular application there are a number of points to be considered.

(a) Reverse Voltage Rating. The maximum voltage that a rectifier will withstand in the reverse direction is known as the 'peak inverse voltage' (p.i.v.). The actual voltage across the rectifier in the reverse direction depends , of course, upon the peak value of the input voltage but also on the type of rectifier circuit used (which is considered in Volume 3 of the series). For the simple circuit of Fig. 2.12 the rectifier must withstand the peak value of the input voltage in the reverse direction. With, say, 240V r.m.s. applied this would be 1·414 × 240 = 339V. Thus a rectifier with a p.i.v. rating of 500V would be suitable for this circuit.

(b) Forward Current Rating. When a semiconductor rectifier is passing current there is a small voltage drop in the forward direction, thus dissipating power that appears as heat in the rectifier. Because of the allowable temperature rise of semiconductor devices there is a limit to the power dissipated. There is, therefore, a limit on the mean forward current passing through the rectifier and also a limit to the permissible maximum peak current. For example, the maximum mean or average (d.c.) current may be 160mA for a silicon diode with a maximum peak current rating of 250mA at an ambient temperature of 25°C. At an ambient temperature of 125°C, these absolute maximum current ratings would be reduced to about 50mA and 125mA respectively. In some rectifier circuits, the peak current passing through the rectifier may be very much larger than the steady current delivered to the load.

(c) Rectifier Capacitance. All semiconductor diodes exhibit capacitance between two terminals, the value depending upon the construcion of the diode. This capacitance becomes important at high frequencies in diodes selected for signal detection in receivers and high speed switching applications. Although the capacitance of semiconductor diodes may be quite high considering their small size, it is not normally important in rectifiers chosen to convert power frequencies from a.c. to d.c.

In addition to observing the above parameters it is often necessary to ensure that excess voltage even of only a few microseconds duration, is not applied to the rectifier. Short duration voltage spikes or transients of considerable magnitude are present in the mains supply and may be generated inside the electronic equipment in which the rectifier is fitted. Protection against these voltage spikes is provided by a capacitor connected in the circuit, usually across the rectifier.

VARACTOR DIODE

When discussing the action of the p-n junction under reverse and forward bias conditions it was noted that the width of the depletion layer varied with the applied voltage. In particular we saw that when the bias was applied in the reverse direction the width of the depletion layer increased causing a rise in the diffusion p.d. The same sort of thing occurs when a voltage is applied to a capacitor. Charges build up on the plates creating a 'back voltage' which opposes the applied voltage. Thus a p-n junction must have a capacitance associated with it.

A p-n junction has a capacitance similar to that of a parallel plate capacitor and the basic idea is shown in Fig. 2.14 where the depletion layer acts as a dielectric and the conductive P and N areas serve as the plates. Fig. 2.14(a) represents the junction in the unbiased state where the distance between the plates is shown as $d1$. As the bias is increased in the reverse direction, the width of the depletion layer is extended and the plate separation increased to $d2$ and $d3$ in Fig. 2.14(b) and (c). Since capacitance is inversely proportional to plate separation, the capacitance of the junction will decrease. Unlike a conventional capacitor, the depletion capacitance varies with the applied voltage and use of this effect is made in Varactor or Variable Capacitance

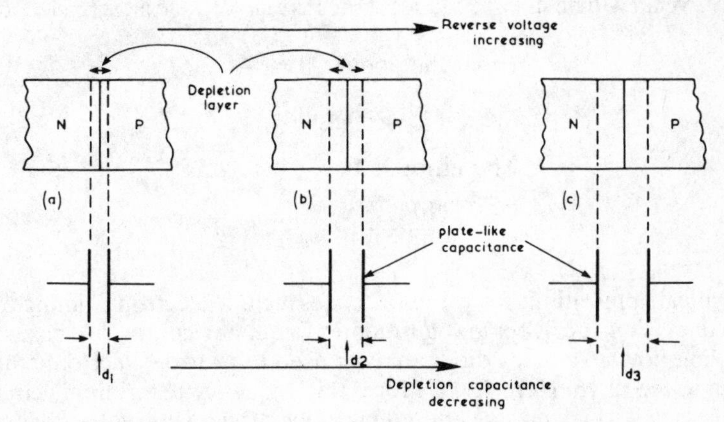

Fig. 2.14 Plate-like capacitance (depletion capacitance) of p-n diode.

(Vari-Cap) diodes. Although all p-n junctions exhibit this variable capacitance effect it is enhanced in varactor diodes by varying the doping levels of the P and N areas. By heavily doping the P-region and lightly doping the N-region, the depletion capacitance lies mainly in the lightly doped side of the junction.

A typical characteristic is given in Fig. 2.15 where the capacitance varies between a maximum of about 23pF to a minimum of 3pF over a reverse voltage range of say, 5–20V. Other varactor diodes may give a capacitance range of 80pF to 25pF or 12pF to 2·5pF to suit different applications.

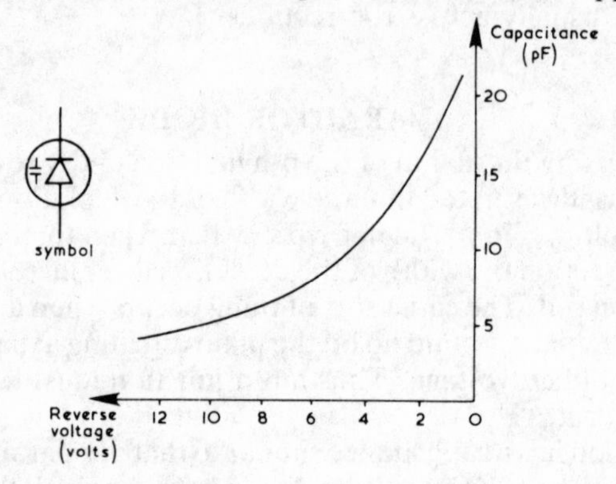

Fig. 2.15 Typical capacitance–voltage characteristic of varactor diode.

The relationship between the capacitance (C) and the reverse voltage (V) is given by

$$C = \frac{k}{\sqrt{V}} \text{ where } k \text{ is a constant}$$

Example: A varactor diode has a capacitance of 78pF at a reverse voltage of 2V. What will be its capacitance if the reverse voltage is increased to 10V?

$$\text{From the above } 78 = \frac{k}{\sqrt{2}}$$

$$\text{or } k = \sqrt{2} \times 78 = 110{\cdot}3.$$

Therefore at 10V

$$C = \frac{110{\cdot}3}{\sqrt{10}} \text{ pF} = 34{\cdot}88 \text{ pF}.$$

Typical applications for varactor diodes include electronic tuning of radio and television receivers and automatic frequency control. A basic circuit showing how a varactor diode may be used to provide electronic tuning is given in Fig. 2.16. Here $L1$, $C1$ form the frequency-determining circuit the frequency of which may be altered by applying a variable voltage from $P1$ to the varactor diode $D1$. As far as the a.c. is concerned, the series combination

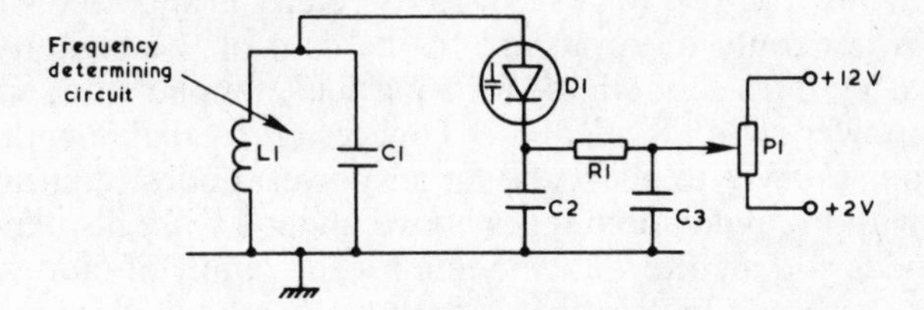

Fig. 2.16 Use of varactor diode for electronic tuning.

of *D*1 and *C*2 are effectively in parallel with the tuning circuit, thus as the reverse bias voltage from *P*1 is altered the capacitance of *D*1 will change so varying the frequency of operation. Further circuit details including varactor diodes are given in Volume 3.

ZENER DIODE

A zener diode is a special silicon junction diode which has a forward characteristic similar to a normal p-n junction diode but is not normally used in the forward direction. In the reverse direction negligible current flows until a certain voltage (called the 'zener voltage') is reached, when the current rises rapidly with little change of voltage across the diode. Operation in this breakdown region does not damage the diode provided the maximum power dissipation for the device is not exceeded. A typical characteristic is given in Fig. 2.17.

The zener or breakdown voltage (which may be altered by varying the impurity levels to set the width of the depletion layer during manufacture) is commonly 2 to 50V, but higher voltage devices are available. Typical zener

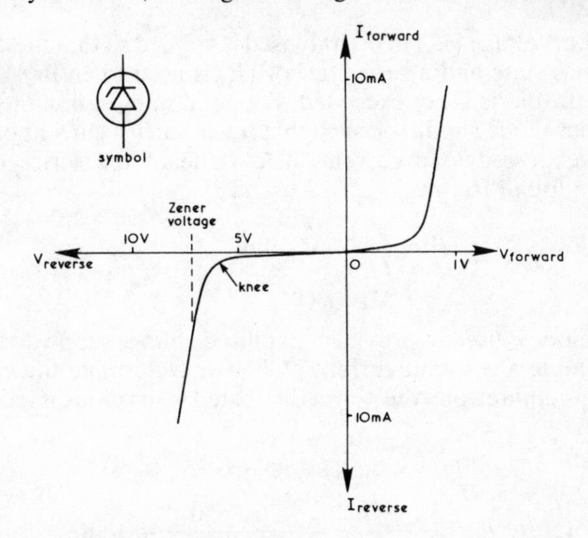

Fig. 2.17 Typical characteristic of zener diode.

voltages are 3·3, 5·6, 6·2, 7·5, 10·6, 12·6, 15·6V, etc. The knee of the reverse characteristic is generally more abrupt in the case of the higher-voltage diodes. When selecting a zener diode for a particular application, one must ensure that its power rating is adequate. Devices are available with power dissipations from 100mW to 50W, the larger power diodes requiring heat sinks. Diodes with breakdown voltages above about 5V are also known as 'avalanche' diodes and an alternative name for the family of diodes where use is made of the breakdown effect is 'voltage regulator diodes'. Below about 5V, the effect (zener) causing the large flow of current is the electric field across the junction which is powerful enough to 'pull' electrons away from the valency bonds of the silicon atoms on the P-side which cross the junction to the N-side These carriers then add to the normal reverse current. Above about 5V, breakdown is due to the 'avalanche effect'. The stronger electric field increases the velocity of the electrons 'pulled' away from the silicon covalent bonds and these electrons collide with atoms, causing ionisation. Extra electrons generated in this way add to the number of carriers crossing the junction and subsequently collide with further atoms producing a cumulative process.

Zener diodes are commonly used to provide constant voltage sources in stabilised power supplies and a basic circuit is shown in Fig. 2.18. The

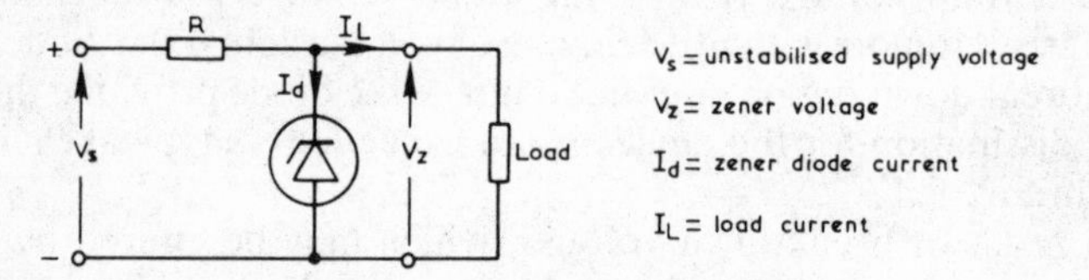

Fig. 2.18 Basic stabilizer using zener diode.

polarity of the d.c. voltage (V_s) to be stabilised is applied so that the diode is in the reverse bias state and a series resitor (R) is used to ensure that the power rating of the diode is not exceeded. For good stabilisation the power rating of the zener diode should be such that it will carry a current of about four times the expected load current. The value of the series resistor required may be found from

$$R = \frac{V_s - V_z}{I_d + I_L} \text{ohm}$$

Example:

A 10V zener diode is used to provide a stabilised voltage supply to feed an amplifier demanding a constant current of 20mA. Determine the value of the series resistor required and the power dissipated in the diode if fed from a 20V supply.

$$I_d = 4 \times I_L = 80\text{mA}$$

$$\text{Therefore } R = \frac{V_s - V_z}{I_d + I_L} = \frac{20 - 10}{100 \times 10^{-3}} = 100\Omega$$

Power (P) dissipated in diode = $V_z \times I_d$ watts = $10 \times 80 \times 10^{-3} = 0{\cdot}8$W.

The following points should be noted in connection with the operation of the basic circuit of Fig. 2.18.

(a) If the load current increases, the diode current will fall in order to maintain a constant voltage drop across R and hence a constant output voltage.

(b) On the other hand if the load current decreases, the diode will pass a larger current to maintain a constant voltage drop across R and hence a constant output voltage.

(c) If the supply voltage (V) is increased, the diode will take a larger current so that extra voltage is dropped across R. Reducing the supply voltage will cause the diode to pass a smaller current so that less voltage is dropped across R thus maintaining a constant output voltage.

Another application for a zener diode is the clipping of various waveforms at a constant voltage level. This use can provide, for example, a waveform of constant amplitude for calibration purposes with the Y-amplifiers of a c.r.o. Further details of circuits using zener diodes are given in Volume 3.

POINT-CONTACT DIODE

A typical construction for a germanium point-contact diode is given in Fig. 2.19. The diode consists of a small pellet of N-type germanium (about 1mm square) attached to one of the connecting leads and a tungsten wire (or whisker) attached to the other connecting lead. The tip of the tungsten wire is made to press against the germanium pellet to form a 'point contact'.

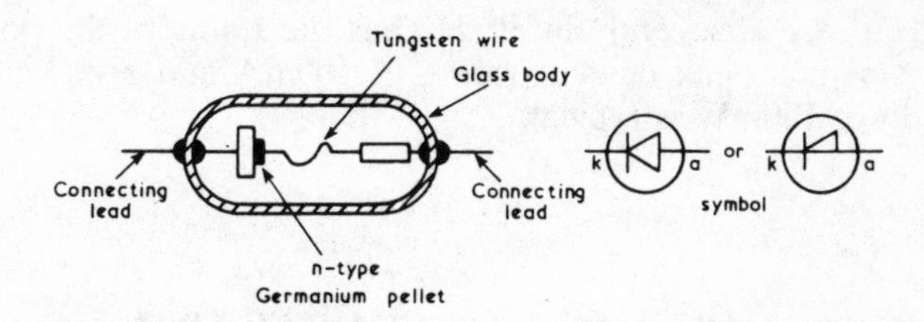

Fig. 2.19 Germanium point-contact diode.

During manufacture, a pulse of current is passed through the diode forming an area of P-type material adjacent to the tungsten wire tip. The diode is then sealed in a small glass bulb for protection against moisture and chemicals.

Apart from its small size this diode has the advantage of a low value of capacitance resulting from the small area of p-n junction that is formed. Due to it small capacitance (0·2 to 1pF) it is commonly used as a signal demodulator in radio and television receivers and may also be used for switching in computer applications.

A typical characteristic is given in Fig. 2.20. In the forward (conducting) direction the voltage drop is small and may be about 1·0V at a current of, say, 4mA rising to about 1·2V at a current of 8mA. In the reverse direction a small reverse current of about 5μA flows with a reverse voltage of 10V rising to 10μA at 40V. At larger reverse voltages the current rises rapidly and a sudden increase occurs if the voltage is increased to a point known as the 'turnover voltage' which varies between about 45V and 150V for the average diode. The maximum reverse voltage applied must be kept well below the turnover voltage.

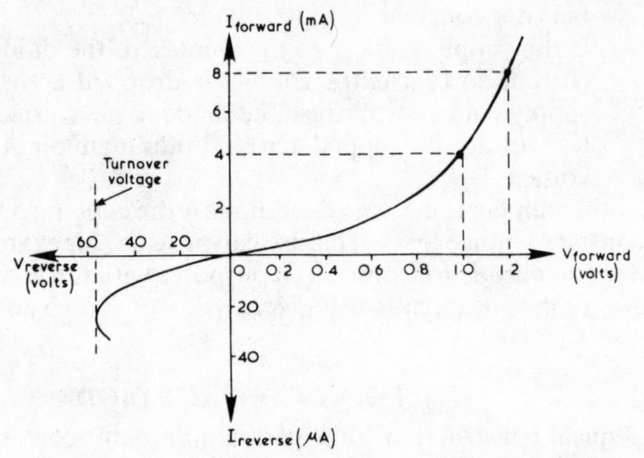

Fig. 2.20 Characteristic of point-contact diode.

Because of the very small area of contact between the whisker and germanium pellet, the heat that can be dissipated without excessive temperature rise is severely limited. Thus the rating of the point contact diode is small: a peak current rating of 100mA and a maximum power dissipation of 100mW is typical.

QUESTIONS ON CHAPTER TWO

(1) A p-n junction is normally made by:
- (a) Forming separate areas of P and N material in a single crystal
- (b) Mixing P and N impurities in a single crystal
- (c) Taking a section of P-type material and joining it to a section of N-type
- (d) Joining a germanium crystal to a silicon one.

(2) The diffusion p.d. is:
- (a) The voltage applied to a p-n junction in the reverse direction
- (b) The voltage applied to a p-n junction in the forward direction

(c) The p-d established across a p-n junction with the P-side positive and the N-side negative
(d) The p.d. established across a p-n junction with the N-side positive and the P-side negative.

(3) The depletion layer of a p-n junction:
(a) Is of constant width
(b) Acts like an insulating zone under reverse bias
(c) Has a width that increases with an increase in forward bias
(d) Is depleted of ions.

(4) Under forward bias conditions the current flowing in the supply leads connected to a p-n diode consists of:
(a) Holes moving from the negative terminal to the positive terminal of the supply
(b) Electrons moving from the negative terminal to the positive terminal of the supply
(c) Holes and electrons moving in opposite directions
(d) Majority carriers only.

(5) The voltage drop across a silicon junction diode with appreciable forward current flow would be about:
(a) 0·2V
(b) 3–4V
(c) 2mV
(d) 800mV.

(6) When maintaining a given forward current in a p-n diode, the voltage drop across the diode will fall for every degree C rise in temperature by:
(a) 0·8V
(b) 40mV
(c) 2mV
(d) 5V.

(7) The maximum voltage that a p-n diode rectifier will withstand in the reverse direction is called the:
(a) Peak-to-peak value
(b) Mean reverse voltage
(c) Zener voltage
(d) Peak inverse voltage.

(8) A point contact diode is normally used for:
(a) Electronic tuning
(b) Power rectification
(c) Signal demodulation
(d) Voltage stabilisation.

(9) A varactor diode may be used for:
  (a) Automatic frequency control in receivers
  (b) Signal demodulation in receivers
  (c) Voltage stabilisation
  (d) Waveform clipping.

(10) A zener diode is normally operated:
  (a) Above its zener voltage
  (b) Below its zener voltage
  (c) In the 'forward' direction only
  (d) In both the 'forward' and 'reverse' directions.

(Answers on page 180)

CHAPTER THREE

THE BIPOLAR TRANSISTOR

THE TERM 'TRANSISTOR' is derived from 'transfer-resistor' and is a device transferring the current in a low-resistance circuit to approximately the same current in a high-resistance circuit. Bipolar means that both holes and electrons are involved in the action of the transistor (as in the p-n diode). A bipolar transistor consists of three separate areas of semiconductor as shown in Fig. 3.1. With two outer P-regions and a central N-region we have a P-N-P transistor as in Fig. 3.1(a); if the areas are given opposite type conductivity as in Fig. 3.1(b) an N-P-N transistor is formed. The two outer regions are called the 'emitter e' and 'collector c' and the central region the 'base b'.

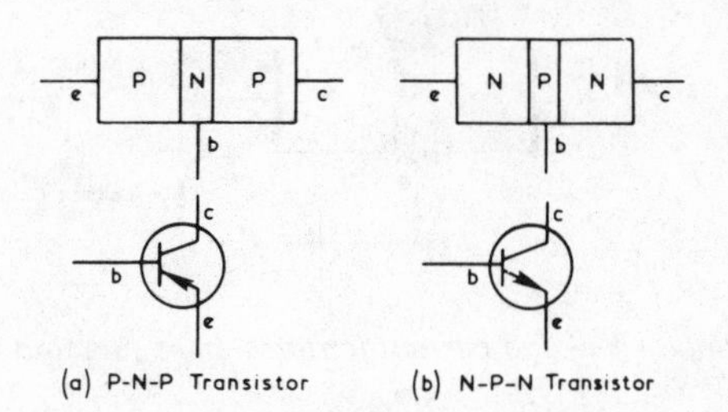

Fig. 3.1 P-N-P and N-P-N bipolar transistors.

Although in this type of diagram emitter and collector regions are shown having the same size this is not so in practice; the collector region is made physically larger as it will normally dissipate the greater power. Both silicon and germanium may be used in the fabrication of the transistor but silicon is more common. The corresponding circuit symbols are given which are similar except for the emitter arrow head. This points in the direction of conventional current flow (or hole movement) through the transistor.

BASIC ACTION

The bipolar transistor consists of two p-n diodes arranged back-to-back thus forming two junctions *J*1 and *J*2. In normal use the emitter–base junction (*J*1) is forward biased and the base–collector junction (*J*2) is

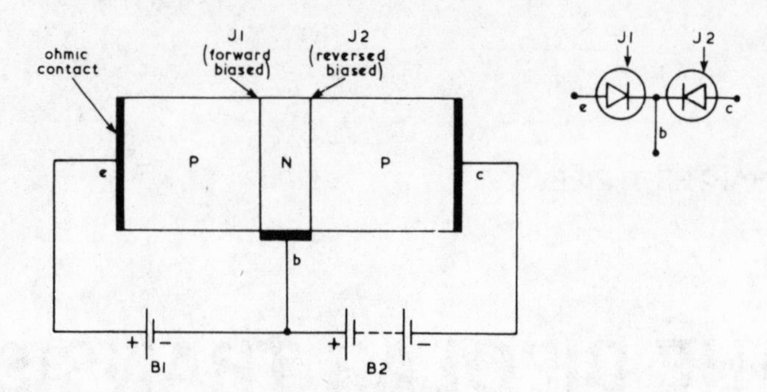

Fig. 3.2 Principle of operation of bipolar transistor (P-N-P).

reverse biased. $B1$ provides the forward bias for $J1$ and $B2$ the reverse bias for $J2$.

Suppose that the bias supply for the emitter–base junction is disconnected as in Fig. 3.3. As the base–collector junction is reverse biased, only a small leakage current will flow between base and collector. This current comprises (in the transistor) minority carriers only; holes moving from base to collector and electrons moving from collector to base regions. If a high resistance load were included in series with the collector lead, the power

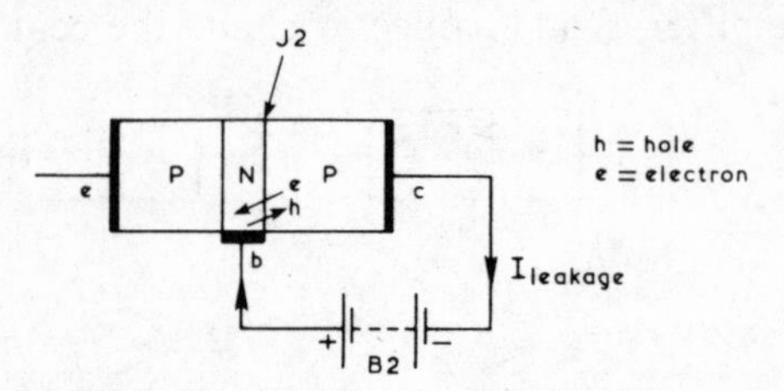

Fig. 3.3 Collector leakage current (emitter–base junction unbiased).

developed would be quite small since the current is very small. Quite high powers could be obtained if extra current carriers could be made available from some other source; this is the purpose of the forward biased emitter–base junction. If the source of the extra carriers is of low power, the device becomes an amplifier, i.e. the output power exceeds the input power.

Consider now Fig. 3.4, where the forward bias supply for the base–emitter junction has been reconnected. With the emitter–base junction in the forward bias state a large current will flow between the emitter and base regions. This current is composed of holes moving from emitter to base and electrons moving from base to emitter. Both are good current carriers, but the only carriers that have a chance of reaching the collector regions are the holes. Thus to ensure that most of the emitter current reaches the collector, the emitter P-region is more heavily doped than the base N-region.

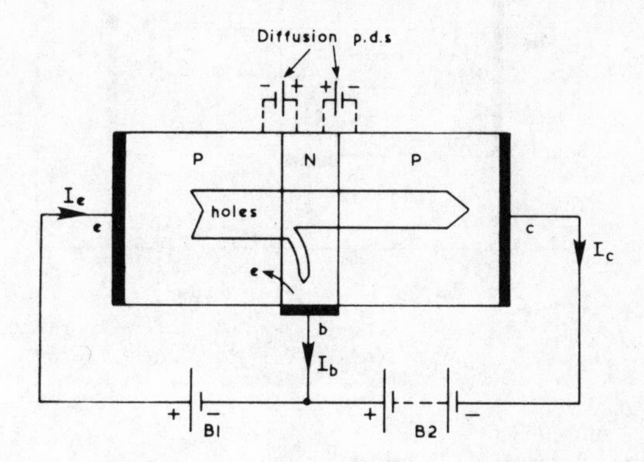

Fig. 3.4 Emitter–base junction forward biased.

Having got a large number of holes into the base region, the problem is to ensure that most of them reach the collector region. Once holes enter the base region they slowly diffuse in all directions. They must not spend too long in the base otherwise they will recombine with electrons and produce base current which is undesirable. Thus the base region must be thin thereby increasing the speed of diffusion through the base. When the diffusing holes reach the base–collector junction they are quickly swept into the collector region by the diffusion p.d. Note that this p.d. across the base–collector junction will aid the passing of holes from the base to collector region, i.e. the injected holes from the emitter appear as minority carriers as far as the base–collector junction is concerned. Most of the emitter current thus reaches the collector and only a small amount arrives at the base. In a well-designed transistor, 0·995 of the emitter current reaches the collector, the remaining 0·005 flowing in the base.

The total current flowing into the transistor must be equal to the total current flowing out, thus

$$I_e = I_c + I_b.$$

It should be noted that the collector current is in two parts: the portion of the emitter current reaching the collector, plus the base–collector leakage current which forms only a small proportion of the total collector current provided the temperature of the device is not excessive.

The action of an N-P-N transistor is just the same as the P-N-P, see Fig. 3.5. Here large numbers of electrons are injected into the base region from the emitter, most of which arrive at the collector. Note that the polarities of $B1$ and $B2$ are reversed and the circuit currents are in the opposite direction.

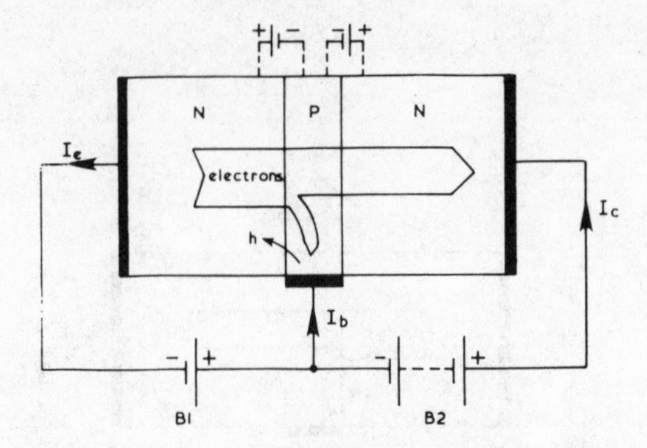

Fig. 3.5 N-P-N transistor action.

Controlling the Collector Current

When a transistor is used as an amplifier it is necessary to be able to vary the collector current, which may be done by altering the emitter current. We have seen that the amount of current flowing across a p-n junction may be altered by varying the forward bias. Thus, in a transistor if the emitter–base forward bias should be increased there will be a larger emitter current and in consequence a larger collector current (also a larger base current). If it is reduced there will be a smaller emitter current and hence a smaller collector current (and smaller base current). This variation in the forward bias of the emitter–base junction is performed by the signal to be amplified as will be seen later. At this stage it should be noted that when the forward bias is altered all three currents are affected – emitter, base and collector. This is important as in some circuits the emitter current is used as the output current.

MODES OF CONNECTION

There are three ways of connecting a bipolar transistor in a circuit. In each, one of the electrodes is common to both the input and output circuits. Although the physical action of the transistor is exactly the same in all of the connections, viewed from the circuit terminals it appears different.

Common Base

In the arrangement shown in Fig. 3.6, the base is common to input and output circuits and is similar to the diagram of Fig. 3.5. Here we are using an N-P-N transistor with *B*2 reverse biasing the biasing the base–collector junction and *B*1 forward biasing the emitter–base junction. E_i represents a sine wave signal source having negligible internal resistance. The currents I_e, I_b and I_c represent the steady or d.c. currents flowing in the emitter, base and collector leads corresponding to the particular forward bias applied

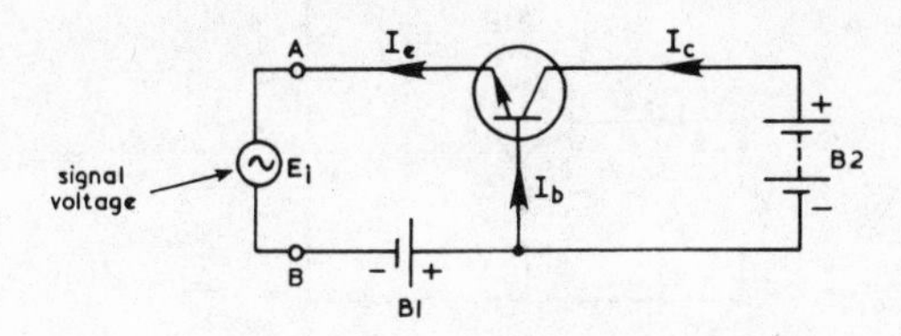

Fig. 3.6 Common-base connection.

from $B1$. In the common-base connection, the **emitter is the input electrode** and the **collector the output electrode**.

Since the signal source is in series with $B1$, any variations for E_i will either add to or subtract from $B1$. Suppose that the signal source is on a positive half-cycle making point A positive with respect to point B. This will reduce the effective emitter–base bias causing the emitter current to decrease and the collector current to decrease. Conversely, when the input signal is on a negative half-cycle making point A negative to point B, the effective emitter–base bias will be increased. This will cause a larger emitter current to flow and hence a larger collector current. In this manner the collector current is made to vary in accordance with the signal voltage E_i.

In a similar way to quoting the resistance value of a resistor or the capacitance value of a capacitor we may quote a figure or figures to indicate particular properties of a transistor. These figures or properties are called 'parameters' and we have already met one for the p-n diode, the a.c. resistance. One important parameter for the common-base circuit is a.c. current gain (h_{fb}) and is defined as:

$$h_{fb} = \frac{\delta I_c}{\delta I_e}$$

with the collector–base voltage held constant (where δ means 'small change of').

This is sometimes called the 'short-circuit forward-current gain' since, with no load in the output circuit, the collector is short-circuited to a.c. by the negligible internal resistance of $B2$. The value of h_{fb} normally lies in the range 0·9 to 0·99 and the closer it is to unity the better.

Common Emitter

The arrangement shown in Fig. 3.7 is that most frequently found in practical circuits and here the emitter is common to input and output circuits. The currents indicated on the diagram are again steady or d.c. components. The base–emitter junction is forward biased by $B1$ and the base–collector junction is reverse biased by the difference in voltage between $B2$ and $B1$. Since $B2$ voltage may be typically 9V and $B1$ voltage small, say, 0·8V for a silicon transistor, the difference is large. In the

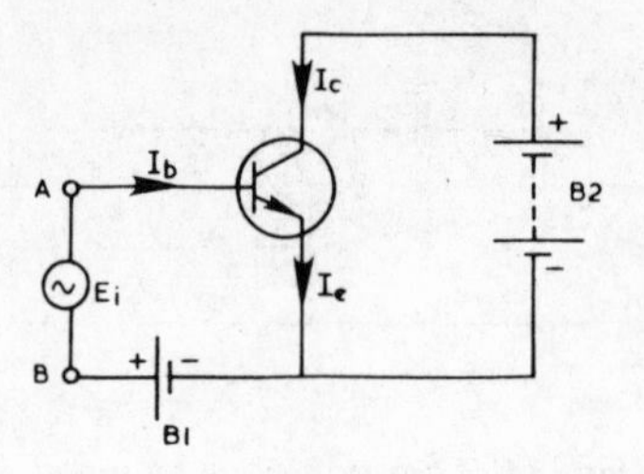

Fig. 3.7 Common-emitter connection.

common-emitter connection the **base in the input electrode** and the **collector the output electrode**.

As before, the signal source E_i is in series with the forward bias supply for the base–emitter junction. Thus when terminal A goes positive with respect to B on the positive half-cycle of the input signal, the forward bias is increased causing the emitter and hence base and collector currents to increase. Conversely, on the negative half-cycle when A goes negative with respect to B the forward bias is reduced causing the emitter and hence base and collector currents to decrease. In this manner the collector current, which is the output current, varies in accordance with the signal voltage E_i. It should be carefully noted, however, that the input current is the base current and not the emitter current as with the common-base connection. This input current, when a signal is applied, must be supplied from the signal voltage source, i.e. a.c. power is required from the signal source.

The a.c. current gain h_{fe} in the common-emitter mode of operation is defined as:

$$h_{fe} = \frac{\delta I_c}{\delta I_b}$$

with the collector–emitter voltage held constant There is a relationship between h_{fe} and h_{fb} which may be seen from the following.

Suppose that the emitter current is made to change by a small amount δI_e causing corresponding small changes in collector current δI_c and base current δI_b. Since the basic current equation holds good we may express this as:

$$\delta I_e = \delta I_c + \delta I_b$$

or when rearranged

$$\delta I_b = \delta I_e - \delta I_c$$

$$\text{Now } h_{fe} = \frac{\delta I_c}{\delta I_b}$$

or substituting for δI_b

$$h_{fe} = \frac{\delta I_c}{\delta I_e - \delta I_c}$$

Dividing numerator and denominator by δI_e gives

$$h_{fe} = \frac{\dfrac{\delta I_c}{\delta I_e}}{1 - \dfrac{\delta I_c}{\delta I_e}}$$

$$\text{but } \frac{\delta I_c}{\delta I_e} = h_{fb}$$

$$\therefore h_{fe} = \frac{h_{fb}}{1 - h_{fb}}$$

Suppose, for example, that h_{fb} is 0·995

$$\text{then } h_{fe} = \frac{0{\cdot}995}{1 - 0{\cdot}995}$$

$$= \frac{0{\cdot}995}{0{\cdot}005} = 199.$$

Thus the a.c. current gain is common emitter can be quite large and may lie in the range 50 to 250.

Common Collector

The third form of connection is shown in Fig. 3.8. Here the collector is common to input and output circuits. The base–collector junction is reverse biased by $B1$ and the base–emitter junction is forward biased by the

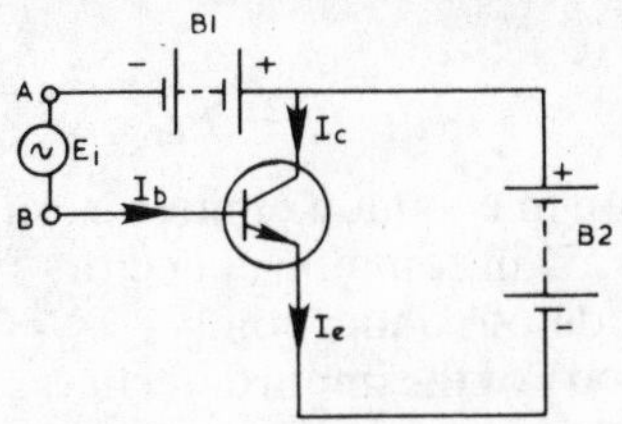

Fig. 3.8 Common-collector connection.

difference between $B2$ and $B1$ voltages. For example, if $B2$ is, say, 9·0V and $B1$ 8·2V, the difference is 0·8V and is of the correct polarity to forward bias the base–emitter junction of an N-P-N silicon transistor. In practice, a resistor would be used in place of $B1$ as will be seen later. In the common-collector connection the **base is the input electrode** and the **emitter the output electrode**.

As with the other connections, the signal voltage E_i will cause variations in the forward bias of the base–emitter junction. On one half-cycle of the input signal when point B goes positive with respect to point A, the forward bias will be increased thus causing an increase in emitter current. On the other half-cycle when B is negative with respect to A, the bias is reduced and the emitter current will decrease. In this manner the emitter output current varies in accordance with the input signal voltage. As with the common-emitter connection the input current is the base current.

The a.c. current gain h_{fc} in the common-collector configuration is defined as:

$$h_{fc} = \frac{\delta I_e}{\delta I_b}$$

with the collector–emitter voltage held constant.

$$\text{Now} \quad \frac{\delta I_e}{\delta I_b} = \frac{\delta I_e}{\delta I_e - \delta I_c}$$

$$\text{or} \quad \frac{\delta I_e}{\delta I_b} = \frac{\dfrac{\delta I_e}{\delta I_e}}{\dfrac{\delta I_e}{\delta I_e} - \dfrac{\delta I_c}{\delta I_e}}$$

(dividing numerator and denominator by δI_e)

$$= \frac{1}{1 - \dfrac{\delta I_c}{\delta I_e}}$$

$$\text{But} \quad \frac{\delta I_c}{\delta I_e} = h_{fb}$$

$$\therefore h_{fc} = \frac{1}{1 - h_{fb}} = h_{fe} + 1$$

If the a.c. current gain in common emitter is, say, in the range 50–250, it will be in the range 51–251 in common collector. Thus there is considerable current gain in this mode of connection.

Table 2 gives a summary of the important characteristics mentioned so far for the three modes of connection

TABLE 2

Mode	**Input Electrode**	**Output Electrode**	**A.C. Current Gain**
Common Base	Emitter	Collector	h_{fb} (0·9 to 0·99)
Common Emitter	Base	Collector	h_{fe} (50 to 250)
Common Collector	Base	Emitter	h_{fc} (50 to 250) approx. same as h_{fe}

CONSTRUCTION OF BIPOLAR TRANSISTORS

There are several methods of manufacturing bipolar transistors but only two will be considered here. One method, used also in the manufacture of junction rectifiers, is shown in Fig. 3.9. Two pellets of indium are alloyed to

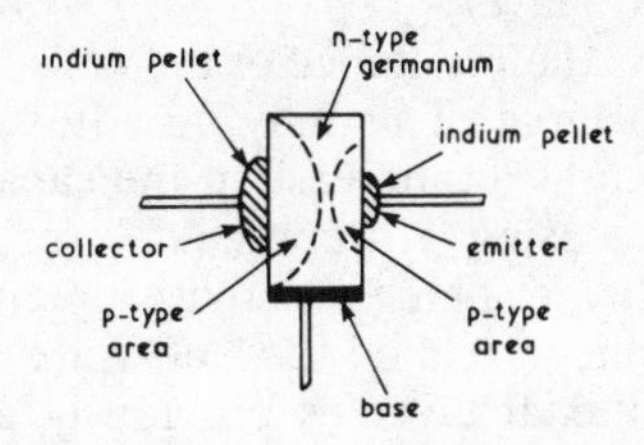

Fig. 3.9 Alloy method of manufacture for p-n-p transistor.

a thin wafer of N-type germanium, resulting in two areas of the germanium slice being converted to P-type. Sandwiched between these areas is a thin region having N-type conductivity which becomes the base layer. It will be noted that the larger P-area forms the collector of the transistor, so emitter and collector connections are not interchangeable as might be suggested by a simple type of schematic diagram such as Fig. 3.2.

An important method of construction is the 'silicon planar' and the various stages in manufacture are given in Fig. 3.10. The starting point is a slice of N-type silicon about 50mm in diameter and 0·3mm thick which is oxidised to a depth of about 1·0μm, see Fig. 3.10(a). A large number of transistors are produced from this slice (about 5000), but only one transistor is considered in the diagram. The next step is to make a 'window' or hole in the oxide layer which is done by photo-etching. A P-type impurity is then allowed to diffuse through the 'window' using a suitable mask to produce the result shown in Fig. 3.10(b). The surface of the wafer is then reoxidised and a smaller window is made through which an N-type impurity is allowed to diffuse to a predetermined depth. This produces the result shown in Fig. 3.10(c). The slice now consists of a P-region sandwiched between two N-regions. Finally, after the surface has been reoxidised and small openings made to take the base and emitter contacts (aluminium), the result is as given in Fig. 3.10(d) with the collector contact, also aluminium.

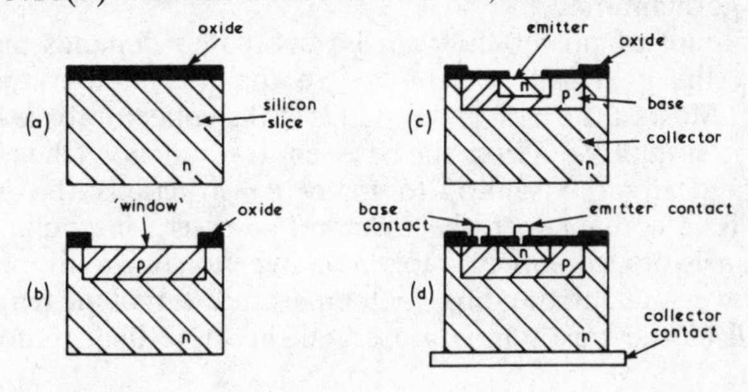

Fig. 3.10 Planar consruction of silicon p-n-p transistors.

The slice is then cut to produce the individual transistors and leads are attached to the base, emitter and collector metal contacts. Each transistor is then encapsulated to keep out moisture and impurities. This type of transistor has the advantage that the top surface (containing the junctions) is protected by the oxide layer.

For some applications the resistance of the collector region in the silicon planar transistor is too large. This may be adjusted without introducing undesirable effects (such as an increase in the capacitance or reduction in breakdown voltage) by using an 'epitaxial layer' in the collector. The epitaxial layer is a thin layer of high resistivity material which is formed on a slice of low resistivity silicon (called the 'substrate'). The transistor is then constructed on the epitaxial layer as previously described to produce a **silicon planar epitaxial transistor**, Fig. 3.11.

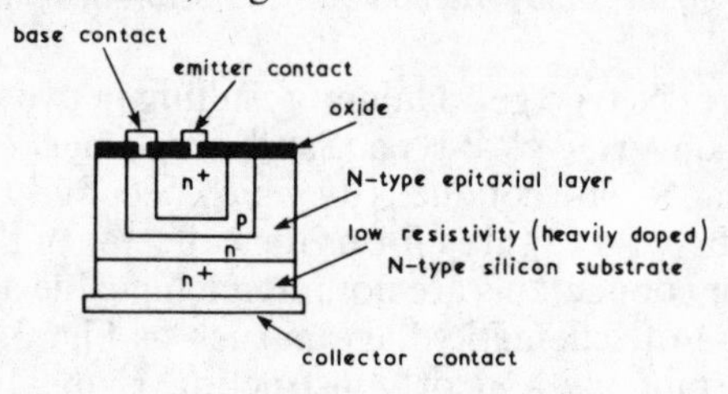

Fig. 3.11 The silicon planar expitaxial transistor (basic structure).

Ratings

Bipolar transistors are made with maximum collector voltage ratings from about 10V to 1kV, which corresponds to the maximum breakdown of the collector–base junction. This voltage should not be exceeded for even short periods or the transistor may be permanently damaged. Maximum current ratings vary from about 10mA to 10A or more and are mainly settled by the maximum power dissipation. When a transistor is operating it has a voltage across it and a current through it, thus there will be a power dissipated equal to $I_c \times V_{ce}$ (assuming only steady current and voltage). This power is converted into heat causing the temperature of the transistor to rise. The maximum temperature is limited to about 150°C for silicon and 85°C for germanium.

The amount of power that can be dissipated depends mainly on how efficiently the heat can be removed so that the temperature rise is not excessive. Most of the heat generated is at the collector–base junction since V_{cb} is greater than V_{be} across the base–emitter junction (the current in each being practically the same). In many constructions the collector–base junction is in contact with the case and so assists in cooling. With small power transistors, cooling is simply by convection due to the air surrounding it. The lower the surrounding air temperature (ambient temperature) the lower will be the case temperature and the more heat removed from the transistor.

To assist in cooling, large power transistors are fitted to a heat sink which is a large mass of metal with a large surface area. The heat sink should be a good heat conductor, e.g. aluminium, and is usually fitted with fins to increase the surface area. Convection, conduction and radiation all play a part in cooling the transistor. If the heat sink is painted matt black the radiation efficiency is increased. Three types of common heat sink construction are shown in Fig. 3.12.

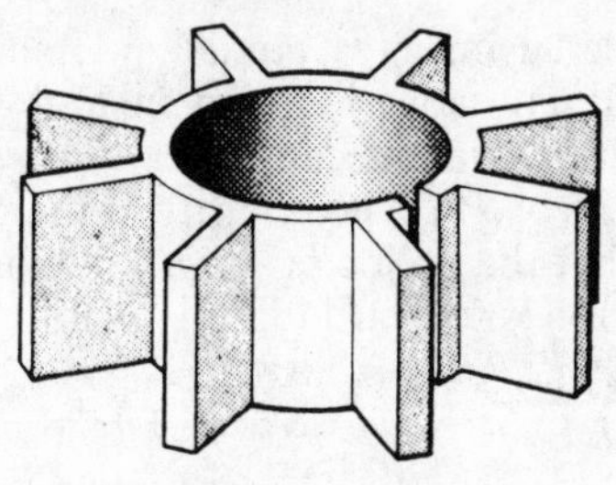

(a) Simple push on heat sink

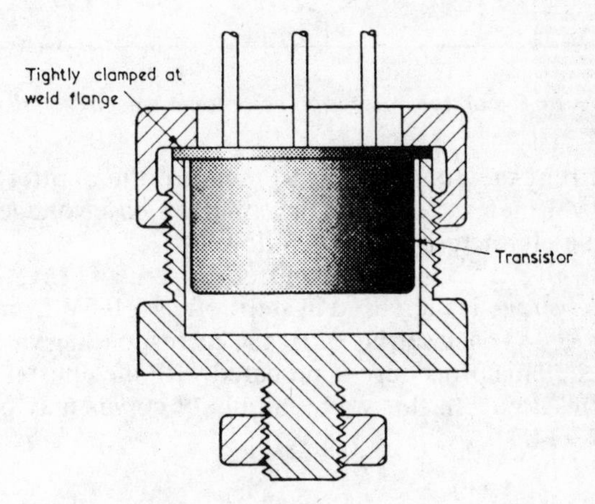

(b) Positive contact type heat sink

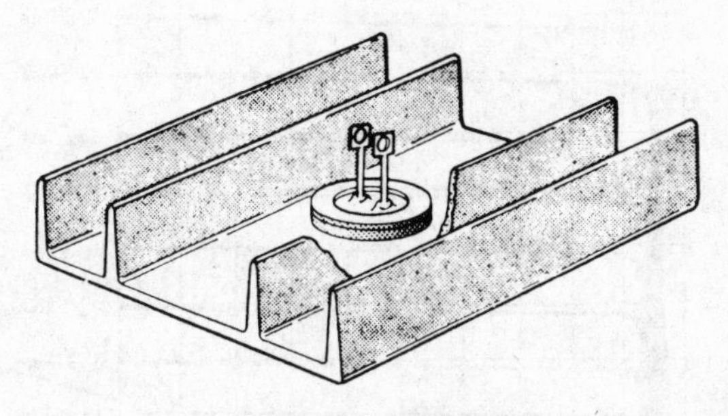

(c) Power heat sink

Fig. 3.12 Heat sinks.

STATIC CHARACTERISTICS

Of the important static characteristic curves for the bipolar transistor only those relating to the common-base and common-emitter connections will be described.

Output Characteristics

These may be obtained using a test circuit and a suitable one for obtaining the common-base characteristics is given in Fig. 3.13. The circuit is so arranged that the emitter current can be varied by means of *R*1 and the collector-to-base voltage by *P*1. The polarity of the voltage supplies *V*1 and *V*2 are connected so that the collector–base junction is reverse biased and the emitter–base junction is forward biased, as for normal operation of the

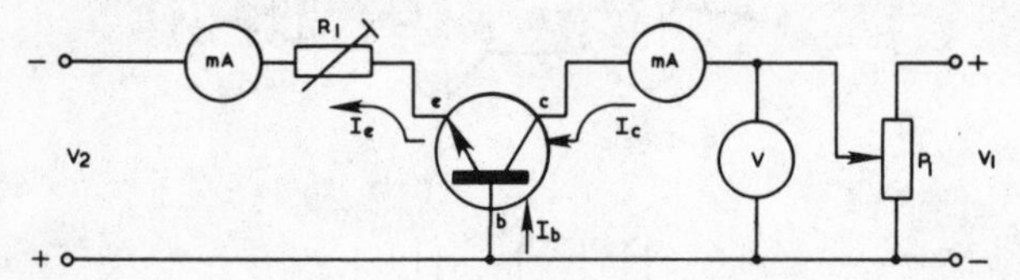

Fig. 3.13 Circuit for measuring common-base characteristics.

transistor. Current meters are used to measure the emitter and collector currents and a voltmeter to indicate the collector–base voltage. The method of obtaining the characteristics is as follows.

*R*1 is set to give a convenient emitter current of, say, 1mA and the collector–base voltage is increased in steps of, say, 0·5V from 0V up to 9V with the aid of *P*1. At each setting of *P*1 the corresponding value of collector current is noted. This procedure is repeated with the emitter current set at 2mA, 3mA, 4mA, etc. In this way a family of curves may be obtained as shown in Fig. 3.14.

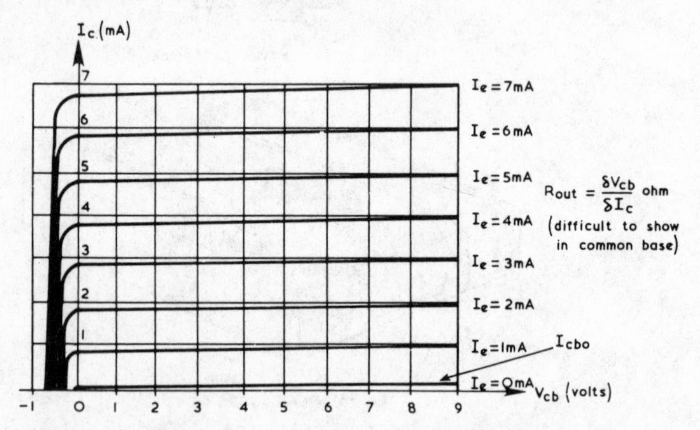

Fig. 3.14 Common-base output characteristics.

The output characteristics of Fig. 3.14 are fairly typical for a small power transistor in common base. It will be noted that the collector–base voltage has negligible effect on collector current, the characteristic being practically horizontal. Another feature is that collector current continues to flow when the collector–base voltage is reduced to zero. This is because the collector–base diffusion p.d. aids the passing of carriers injected from the emitter. Thus, to reduce the collector current to zero, the diffusion p.d. must be reduced to zero. This is achieved by reversing the collector–base voltage to a few tenths of a volt. These characteristics and others that follow are called 'static characteristics' because the collector voltage is held steady when the collector current changes, (unlike a normal amplifier when a load is included in the collector circuit).

Another important parameter of the transistor is the a.c. output resistance (R_{OUT}). This is defined as:

$$R_{OUT} = \frac{\delta V_{cb}}{\delta I_c} \text{ ohm (with } I_e \text{ constant).}$$

Since the characteristics are nearly horizontal (small slope), the output resistance is high and of the order of 100kΩ to 1MΩ.

Common-emitter output characteristics are given in Fig. 3.15. These may be obtained using a type of test circuit similar to Fig. 3.13. This time, however, the base current is set to a convenient level, say 25μA and the collector–emitter voltage increased in discrete steps. At each step the corresponding value of collector current is noted. This is repeated for various values of base current so building up a family of characteristics.

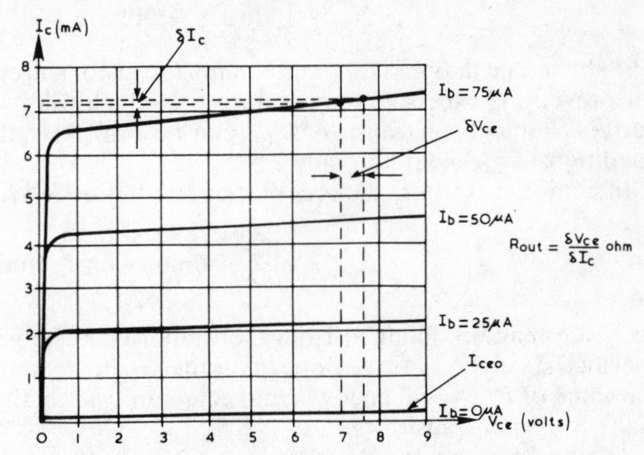

Fig 3.15 Common-emitter output characteristics.

Although in a general way the common-emitter output characteristics are similar to those in common base there are a number of differences. First, the collector–emitter voltage has a greater influence on the collector current, i.e. the characteristics are not so horizontal as in Fig. 3.14. This is because the common-emitter d.c. current gain increases by a greater amount with an increase in collector–emitter voltage than the d.c. current gain in common base. Secondly, the characteristics are not as parallel which again is due to the larger change in common-emitter d.c. current gain than common-base d.c. current gain at the higher currents. Thirdly, when V_{ce} is zero then the collector current has fallen to zero irrespective of the value of base current. This is because once the collector–emitter voltage has fallen below the base–emitter voltage (say 0·6V for a silicon transistor), the **collector–base** junction becomes **forward biased** and not reverse biased. The collector current remains at full level until the collector–emitter voltage falls to about 0·3V below that of the base–emitter voltage at which point the collector current falls abruptly (the forward bias of the collector–base junction prevents charges injected from the emitter reaching the collector). This minimum value of collector–emitter voltage when maintaining full collector current is called the 'bottoming voltage' and will be considered again in Chapter 5. Finally, with zero base current there is still some collector current flowing.

This is the common-emitter leakage current I_{CEO} and is greater than the leakage current I_{CBO} in common base (see Fig. 3.14). The leakage currents are related by the expression:

$$I_{CEO} = I_{CBO}\,(1 + h_{fe}).$$

Assuming a leakage current of, say, 2μA in common base, it will be in common emitter (with an h_{fe} of 100)

$$2(1 + 100)\ \mu\text{A} = 202\mu\text{A}.$$

The above calculation is for a germanium transistor and emphasises the need for preventing excessive temperature rise which will increase the leakage current. For a silicon transistor I_{CBO} may be only 25nA, thus under the same conditions I_{CEO} would be only 2·525μA.

In common emitter, the a.c. output resistance (R_{OUT}) is defined as

$$R_{OUT} = \frac{\delta V_{ce}}{\delta I_c} \text{ ohm (with } I_b \text{ constant)}$$

Its value may be obtained from the output characteristic as shown for a specified steady V_{ce}. As the slope of the lines is greater than in common base, the value of R_{OUT} will be lower and commonly lies in the range of 10kΩ to 50kΩ for small transistors.

The common-emitter output characteristics of a high-power transistor capable of providing up to 100W for industrial electronics applications is given in Fig. 3.16 for comparison with Fig. 3.15. Note that I_c is in amperes and I_b correspondingly larger in milliamperes The a.c. output resistance for a transistor of this type will be of the order of 5–10 ohms. Power transistors with ratings somewhat lower, about 30W, are used in colour television receivers and audio equipment.

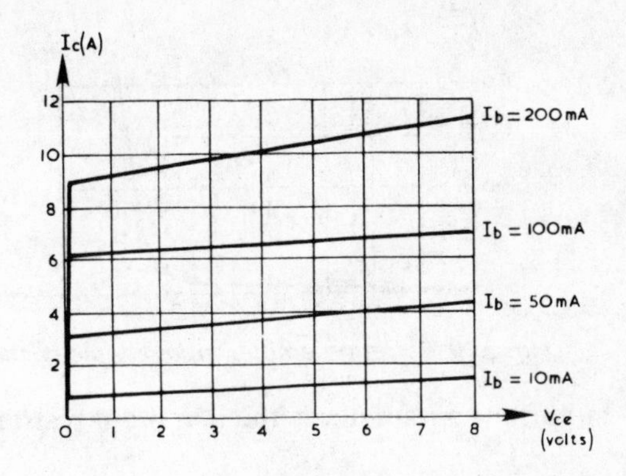

Fig. 3.16 Common-emitter output characteristic for large power transistor.

Input Characteristics

Another useful characteristic is the input characteristic which shows how the input current varies with the voltage applied between base and emitter. A typical input characteristic for the common-base connection is shown in Fig. 3.17. As the forward bias is increased from zero there is practically no emitter current until V_{cb} exceeds about 0·5V for the silicon transistor shown, or about 0·1V for a germanium type. Thereafter, the current increases abruptly in a non-linear manner. The non-linearity of the characteristic will give rise to distortion when the transistor is used as an amplifier but this may be reduced as explained later.

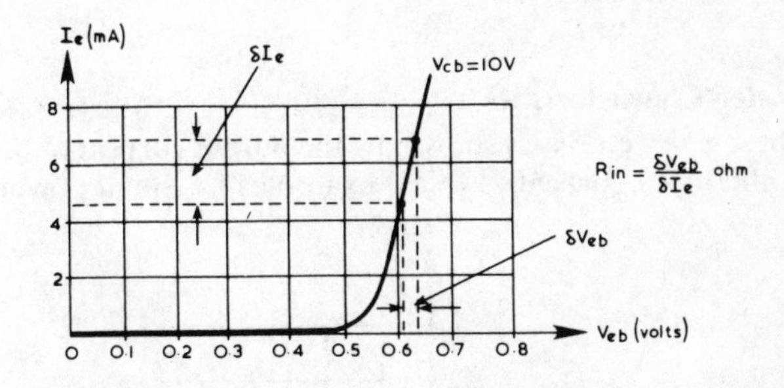

Fig. 3.17 Common-base input characteristic.

A typical input characteristic for the common-emitter connection is given in Fig. 3.18. This is similar to the common-base characteristic except that in common-emitter the input current is the base current. Again, the input current rises abruptly when the forward bias reaches about 0·6V for a silicon transistor.

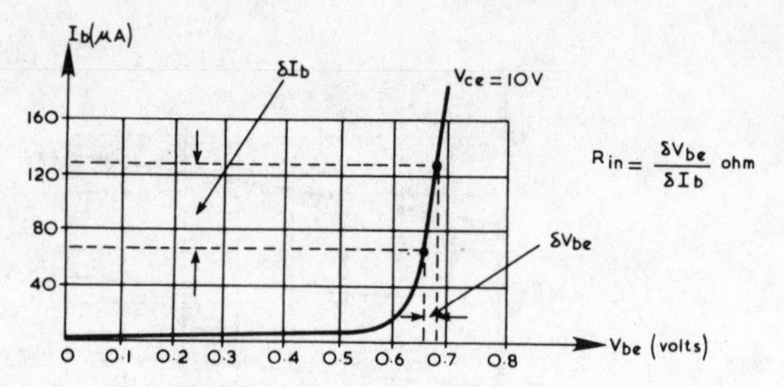

Fig. 3.18 Common-emitter input characteristic.

A further important parameter is the a.c. input resistance (R_{IN}). This is defined as:

For common base

$$R_{IN} = \frac{\delta V_{eb}}{\delta I_e} \text{ ohm } (V_{cb} \text{ constant})$$

and for common emitter

$$R_{IN} = \frac{\delta V_{be}}{\delta I_b} \text{ ohm } (V_{ce} \text{ constant})$$

The values may be obtained from the input characteristics by considering small changes as shown. The value will depend upon the point of measurement since the curve is non-linear. In common-base, R_{IN} lies in the range of about 30Ω–100Ω whereas in common-emitter it is higher, lying in the range of about 750Ω–5kΩ. These values are for small power transistors. A high-power transistor operating in common-emitter may have an R_{IN} of only 0·5Ω to 10Ω.

Transfer Characteristics

These characteristics show how the **output current** of the transistor varies with the **input current**. Typical examples for small power transistors when

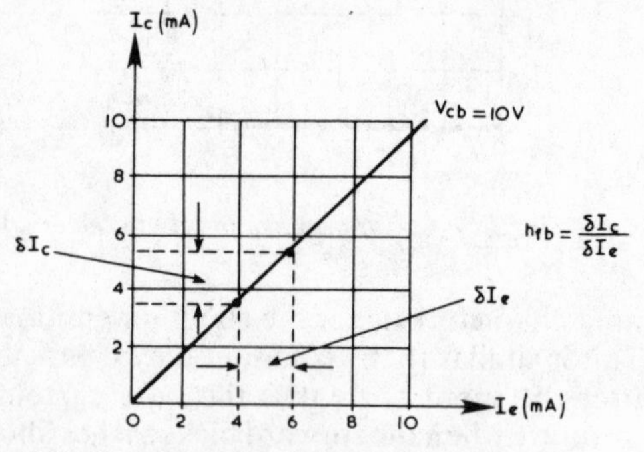

Fig. 3.19 Common-base transfer characteristic.

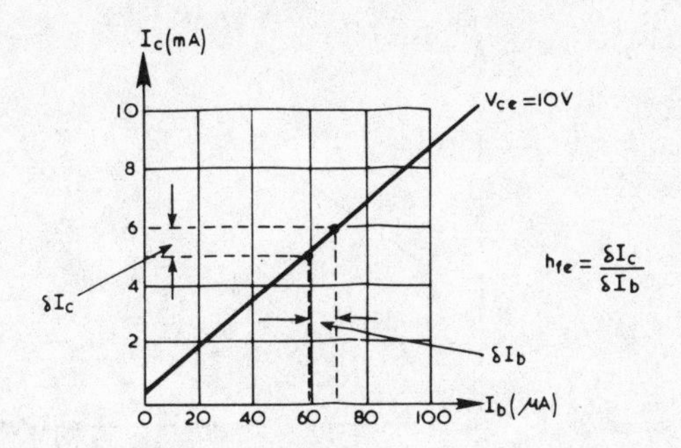

Fig. 3.20 Common-emitter transfer characteristic.

operating in common-base and common-emitter are given in Figs. 3.19 and 3.20 respectively.

The transfer characteristic may be used to determine the value of h_{fb} or h_{fe} by considering small changes as illustrated. Departure from the linear characteristic shown often occurs at low and high values for input current.

Frequency and Gain Characteristics

The current gain h_{fb} or h_{fe} does not remain constant at all frequencies, as time is required for carriers injected from the emitter to diffuse through the base region, which is a relatively slow process. Thus the output is slightly delayed on the input which cause a reduction in gain at high frequencies. There is, therefore, an upper limit on the useful frequency range of any transistor. This upper frequency is called the 'cut-off frequency' ($f_β$) and is defined as the frequency where the gain of the transistor has fallen to 0·707 of its low-frequency value, see Fig. 3.21. For a transistor used in the audio output stage of a radio receiver, $f_β$ should be at least 20kHz whereas in the video stage of a t.v. receiver an $f_β$ of 6MHz would be suitable.

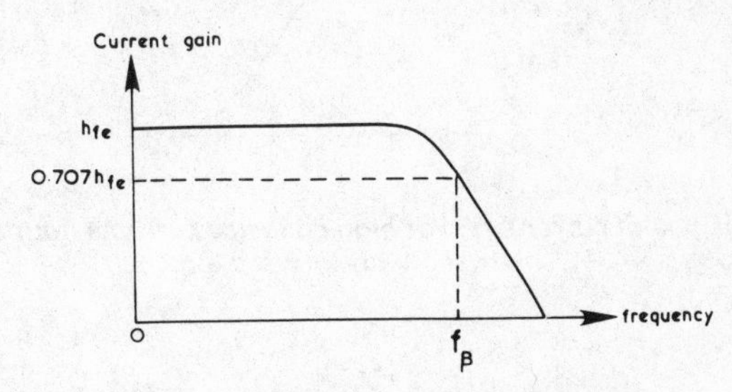

Fig 3.21 Variation of current gain with frequency.

The manner in which the current gain varies with collector current is important and for many transistors the relation is similar to that shown in Fig. 3.22. Generally, as the collector current (or emitter current) is increased the value of h_{fe} increases as shown. Use of this effect is made in receivers

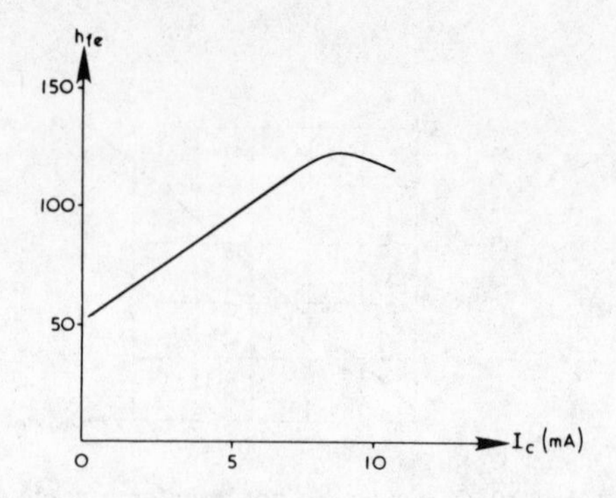

Fig. 3.22 Variation of current gain with collector current (found in transistors used with reverse a.g.c.).

fitted with automatic gain control (a.g.c.). By reducing the forward bias, thereby decreasing the collector current, the gain of the transistor may be reduced. This method of reducing gain is known as 'reverse a.g.c.'.

Some transistors are specially designed to exhibit a collapse in current gain when the collector current rises above a high value, say 5mA. Fig. 3.23 shows the characteristic of such a transistor. This type is used in high-frequency receivers, e.g. f.m. radios and t.v. receivers, for a.g.c. purposes. By increasing the forward bias the collector current increases causing a reduction in gain. This method of reducing transistor gain is called 'forward a.g.c.'

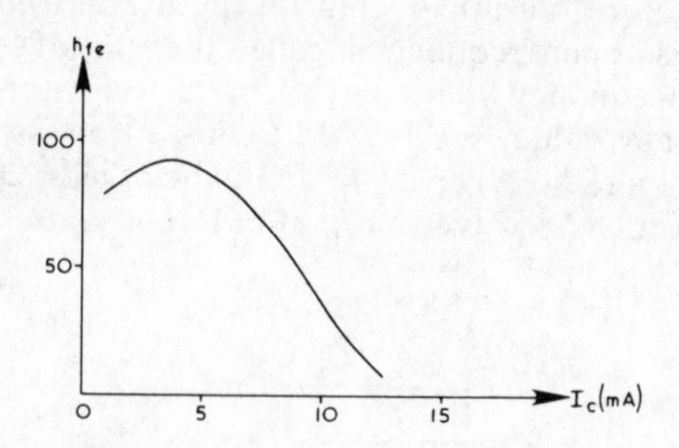

Fig. 3.23 Collapse of current gain at high collector currents (found in some transistors used with forward a.g.c.).

QUESTIONS ON CHAPTER THREE

(1) In a bipolar transistor:
(a) The collector–base junction is forward biased and the emitter–base junction reverse biased
(b) The collector–base junction is reverse biased and the emitter–base junction forward biased
(c) The emitter–base junction and the collector–base junction are both reverse biased
(d) The emitter–base junction and the collector–base junction are both forward biased.

(2) With an N-P-N transistor:
(a) Emitter and collector are normally positive to the base
(b) Emitter and base are normally positive to the collector
(c) Collector and base are normally negative to the emitter
(d) Collector and base are normally positive to the emitter.

(3) The relation between the collector, base and emitter currrents is given by:
(a) $I_c = I_e + I_b$
(b) $I_e = I_b + I_c$
(c) $I_b = I_c - I_e$
(d) $I_e = I_c - I_b$.

(4) A typical value for the h_{fb} of a transistor would be:
(a) 0·99
(b) 50kΩ
(c) 100Ω
(d) 50.

(5) The h_{fe} of a transistor may be defined as:
(a) $\frac{\delta V_{ce}}{\delta I_c}$
(b) $\frac{\delta I_c}{\delta V_{ce}}$
(c) $\frac{\delta I_b}{\delta I_c}$
(d) $\frac{\delta I_c}{\delta I_b}$.

(6) The output current in common collector is taken from:
(a) The collector
(b) The base
(c) The emitter
(d) The signal source.

(7) The a.c. output resistance of a small power transistor in common-emitter is about:
(a) 10–50Ω
(b) 10kΩ–50kΩ
(c) 200Ω–1kΩ
(d) 200kΩ–2MΩ.

(8) The a.c. input resistance for a small power transistor connected in common-base is about:
(a) 750Ω–5kΩ
(b) 7·5kΩ–50kΩ
(c) 30Ω–100Ω
(d) 3kΩ–100kΩ.

(9) The a.c. input resistance of a transistor connected in common-base is lower than that in common-emitter because:
(a) The physical action of the transistor is different
(b) The input current is greater for similar changes in V_{be}
(c) The output current is greater for similar changes in V_{be}
(d) The input current is practically zero.

(10) The transfer characteristic of a bipolar transistor shows the relationship between:
(a) V_{be} and output current
(b) V_{be} and input current
(c) V_{be} and V_{ce}
(d) Input current and output current.

(11) The input characteristic of a transistor in common-base shows the relationship between:
(a) V_{be} and I_e
(b) V_{be} and I_b
(c) I_c and I_b
(d) V_{cb} and V_{be}.

(12) For most transistors the current gain:
(a) Is constant with changes in I_c
(b) Increases with a decrease in I_c
(c) Decreases with an increase in I_b
(d) Increases with an increase in I_c.

(Answers on page 180)

CHAPTER FOUR

THE FIELD-EFFECT TRANSISTOR

ALTHOUGH THE PRINCIPLE of the field-effect transistor (fet) has been known for quite some time, it is only in recent years that semiconductor technology has made possible its production on a commercial scale.

An fet is a resistive channel of semiconductor material (silicon) whose resistance can be varied by an electric field cutting into the channel and altering its cross-sectional area. The fet is a **unipolar** device, i.e. it conducts by majority carriers only, a feature that makes it less temperature dependent. There are a number of types available which can be either conducting or non-conducting with zero bias applied.

JUNCTION GATE FET (JUGFET)

The basic idea of the junction gate fet is given in Fig. 4.1. It consists of a thin wafer of either N-type or P-type silicon with ohmic contacts at each end called the 'source' (s) and 'drain' (d). On opposite faces there are diffused two heavily doped areas which are connected together. These areas are P-type with an N-channel device and N-type for a P-channel device. The areas form the 'gate' (g) or control electrode. Source, gate and drain electrodes correspond to the emitter, base and collector connections of the bipolar transistor. The arrow in the circuit symbol points in the direction of

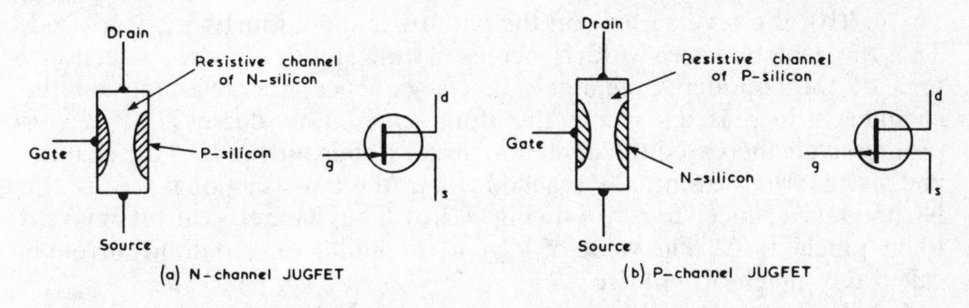

Fig. 4.1 Basic form of construction for junction-gate fet.

forward current flow in the p-n junction formed between the gate and the resistive channel.

If we consider an N-channel jugfet as in Fig. 4.2(a) where the drain is held positive with respect to the source, then electrons (majority carriers) will flow in the resistive channel between source and drain. With zero voltage between the gate and source ($V_{gs} = 0V$), i.e. the gate at the same potential as the source, a reverse biased junction is formed between the gate areas and the resistive channel. When a voltage is applied between drain and source there is a voltage drop along the channel. Thus the portion of the N-channel close to the P-area will be positive to the source by an amount depending upon the manufactured dimensions of the P-areas, which are different in practice from the simple idea shown in Fig. 4.2. As with any p-n junction there will be a depletion layer and let us say that under the bias conditions stated it has the area shown. This layer will lie mainly in the lightly doped N-channel and will be wider at the top of the P-area than at the bottom since the reverse bias is greater towards the top of the P-area. Under these conditions maximum current flows in the resistive channel and the drain and source connections, say 10mA. Since the gate–channel junction is reversed biased, only a very small leakage current will flow in the gate lead.

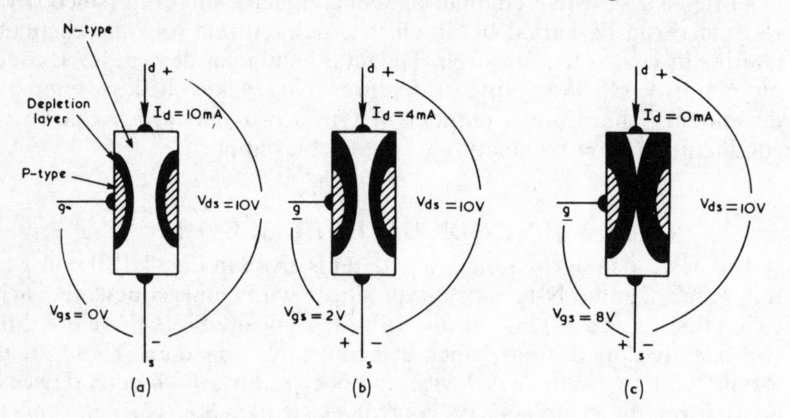

Fig. 4.2 Effect of gate–source voltage on drain current.

If the gate is made negative with respect to the source by, say, 2V as in Fig. 4.2(b), the reverse bias on the gate-to-channel junction is increased. Thus the depletion area width is increased which reduces the cross-sectional area of the conductive channel. In consequence, the resistance of the channel is increased causing the drain current to decrease. If V_{gs} is progressively increased the depletion area extends further into the channel and a situation is eventually reached where the cross-sectional area of the N-channel is reduced to zero as in Fig. 4.2(c). The channel is cut-off or is said to be 'pinched-off'. The value of V_{gs} corresponding to zero drain current is called the 'pinch-off voltage'.

A typical transfer characteristic (or mutual characteristics) of an N-channel jugfet is given in Fig. 4.3(a) which shows how the drain current varies with the bias voltage applied between gate and source. The characteristic is not linear but follows an approximate square law. There is a maximum bias voltage that may be applied between gate and source, this being the p-n junction breakdown voltage. If the gate is made positive to the source, the drain current will increase. However, if made too positive the gate-channel diode will become forward biased. The gate is not usually operated with a positive gate voltage above about 0·5V.

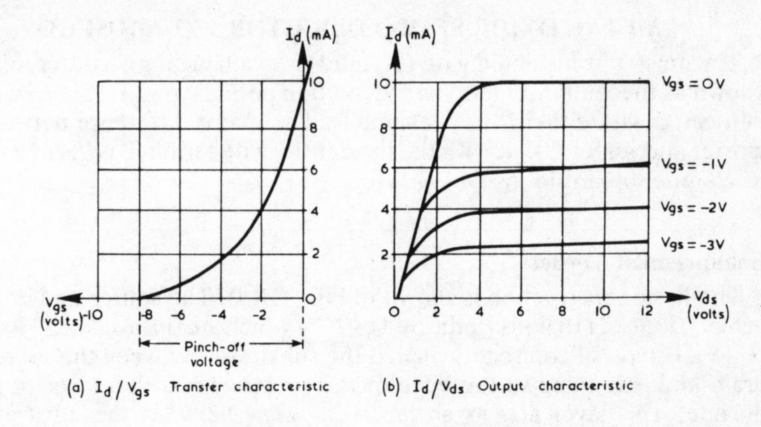

Fig. 4.3 Characteristic of jugfet (common source).

A family of output characteristics is given in Fig 4.3(b). With a fixed gate voltage the current increases fairly linearly at first with increases in the drain-to-source voltage. Thereafter, the drain current is almost independent of V_{ds}.

Parameters

The fet is a voltage-operated device, unlike the bipolar transistor which is current operated. One important parameter of the fet is the mutual conductance (g_m).

This is defined as $g_m = \dfrac{\delta I_d}{\delta V_{gs}}$, with V_{ds} maintained constant.

As this is an I/V relationship it is measured in the unit of conductance, the siemens (S). This parameter shows how effective are changes in V_{gs} in producing variations in the drain current. Typical values for g_m lie in the range 1·5mS to 7mS.

Other useful parameters are the a.c. input resistance R_{IN} and the a.c. output resistance R_{OUT}.

R_{IN} may be defined as $\dfrac{\delta V_{gs}}{\delta I_g}$ ohm, with V_{ds} constant.

Since the input is a reversed biased juction the gate current (I_g) is extremely small, typically of the order of tens of pico-amps. Thus R_{IN} is very high, about 10^7 ohms at low frequencies.

$$R_{OUT} \text{ may be defined as } \frac{\delta V_{ds}}{\delta I_d} \text{ ohm, with } V_{gs} \text{ constant.}$$

The value of R_{OUT} is normally very high, lying in the range 20kΩ to 1MΩ.

METAL OXIDE SEMICONDUCTOR FET (MOSFET)

The mosfet or insulated gate fet (igfet) is available in two different forms known as the 'enhancement' and 'depletion' mode types. Both types can be fabricated with either P or N channels. The major difference between the mosfet and jugfet is that with the mosfet the gate terminal is insulated from its channel by a thin insulating layer.

Enhancement Mosfet

The basic construction is shown in Fig. 4.4(a). The source and drain are heavily doped N regions (indicated by N^+) which are diffused close together on to a P-type silicon section called the substrate. Between the source and drain and extending to the edge of the substrate is a thin layer of silicon dioxide. This layer acts as an insulating zone between the substrate and aluminium gate electrode. A connection is also made to the substrate and in most transistors it is internally connected to the source terminal but may be brought out as a separate electrode.

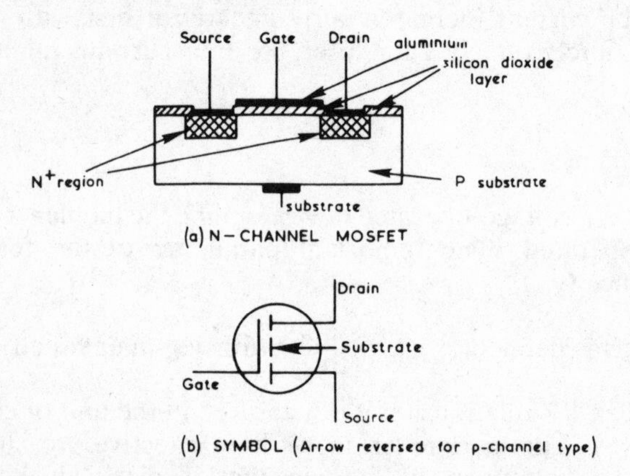

Fig. 4.4 Mosfet (enhancement type).

With the drain made positive to the source and the gate voltage zero, very little current flows since the two p-n junctions formed are in series opposition. If the gate is made positive to the source, the electric field set up across the silicon dioxide layer attracts electrons out of the substrate. This creates an n-type surface channel between the source and drain through which conduction may take place. Hence the enhancement mosfet has to be forward biased like the bipolar transistor, but of course is voltage operated.

When the gate-to-source voltage is increased in the positive direction as in Fig. 4.5, the width of the surface enhancement channel is increased. As a result the resistance of the channel is decreased thereby permitting a larger

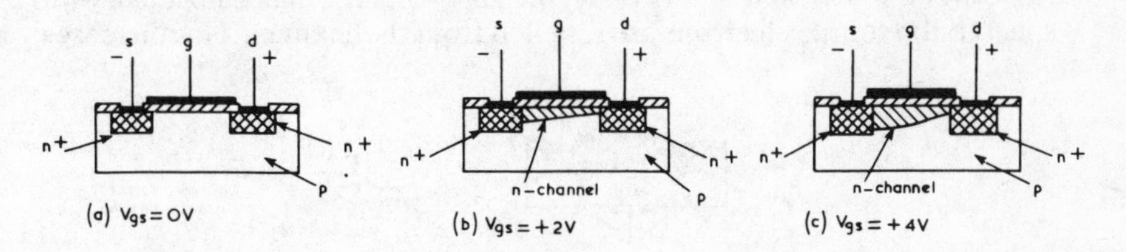

Fig. 4.5 Effect of forward biasing the gate.

current to flow between source and drain. Typical characteristics are given in Fig. 4.6. Drain current does not commence to flow until the gate is made slightly positive with respect to the source; the gate voltage at which drain current commences to flow is called the 'threshold voltage' (V_{th}).

Owing to the presence of the silicon dioxide layer, the input resistance is higher than the jugfet ($10^{10}\Omega$ or more). To prevent the possibility of damage to the gate oxide layer by an electrostatic charge building up on the high resistance gate electrode, the mosfet is normally supplied with a conductive rubber ring fitted round the leads of the device. This ring should not be removed until after the transistor has been mounted in the circuit.

An n-channel device has been considered here but a p-channel mosfet may be formed using an N-type substrate and highly doped P-areas. Conduction through the channel is enhanced by making the gate negative with respect to the source.

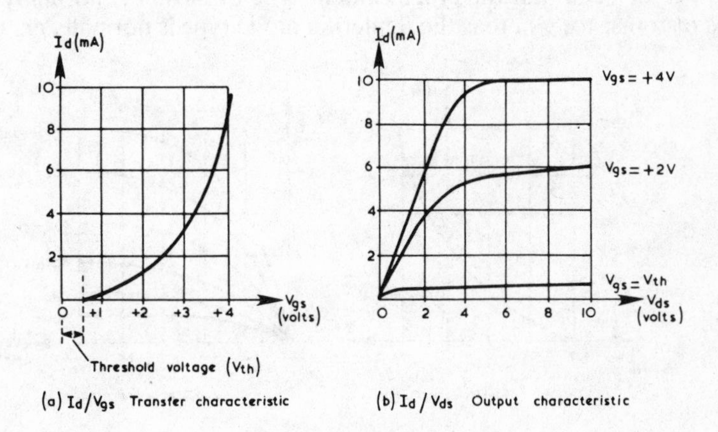

Fig. 4.6 Characteristics of enhancement type mosfet.

Depletion Mosfet

The principle of operation of the depletion mode mosfet is illustrated in Fig. 4.7, using an n-channel device as an example. Between the highly doped source and drain regions a small n-channel is formed during manufacture. This channel has a higher resistivity than the n^+ sections. If the drain is held at a positive voltage with respect to the source and V_{gs} is zero, a current will flow between source and drain. When the gate voltage is made negative with respect to the source electrons are repelled from the channel. This increases

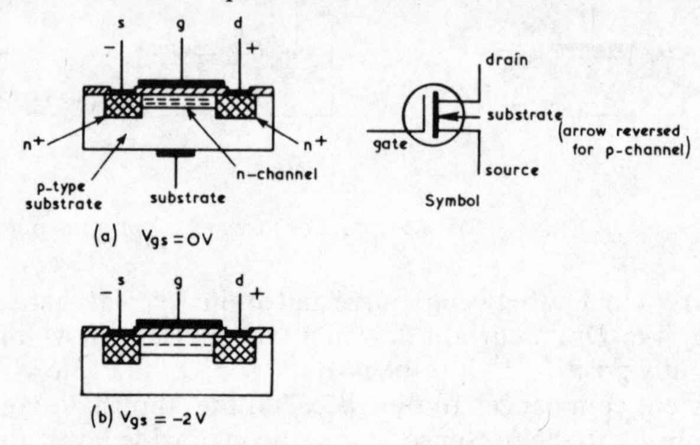

Fig. 4.7 Depletion mode mosfet (n-channel).

the resistivity of the channel thereby reducing current flow. The more negative the gate voltage, the more depleted the channel becomes of electrons thus the less the current flow, Fig. 4.7(b). Typical characteristics are given in Fig. 4.8. If the gate is made positive with respect to the source, electrons will be attracted into the channel and the source-to-drain current increased. Thus, over the positive range of gate voltage this mosfet behaves as an enhancement type.

It will be seen that the enhancement type of mosfet is normally an 'off' type of transistor whereas the depletion mode type is normally an 'on' type

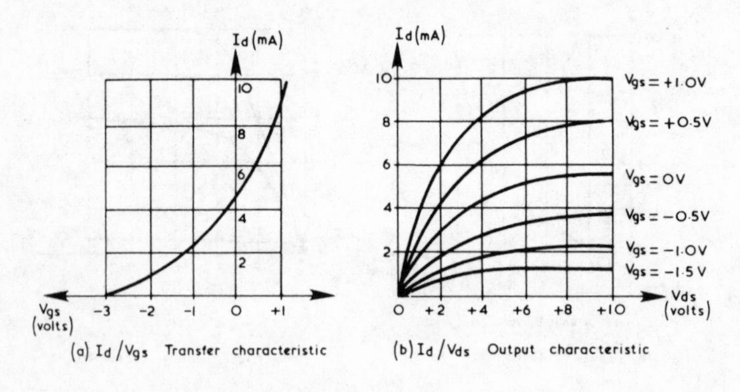

Fig. 4.8 Characteristics of depletion-type mosfet.

of device. In either case the current flowing between source and drain can be varied by altering the gate-to-source voltage.

The main parameters previously discussed for the junction-gate fet apply also to the mosfet device. The parameter g_m is typically in the range 2–10mS; R_{IN} lies in the range 10^{10}–10^{12} ohms and R_{OUT} typically in the range 10–50kΩ.

Circuit Configurations of the Fet

In the same way that bipolar transistors may be connected in three different configurations (as explained in Chapter 3), fets can be connected in:

(a) The common-source mode (corresponding to common emitter);
(b) The common-drain mode (corresponding to common collector);
(c) The common-gate mode (corresponding to common base).

The basic circuit configurations using a jugfet as an example are shown in Fig. 4.9. A battery $B2$ is used to provide the correct biasing of the fet, but in practice the bias would be obtained from a resistor connected in the source circuit which will be described later. All three connections may be used but the common source arrangement is the most usual.

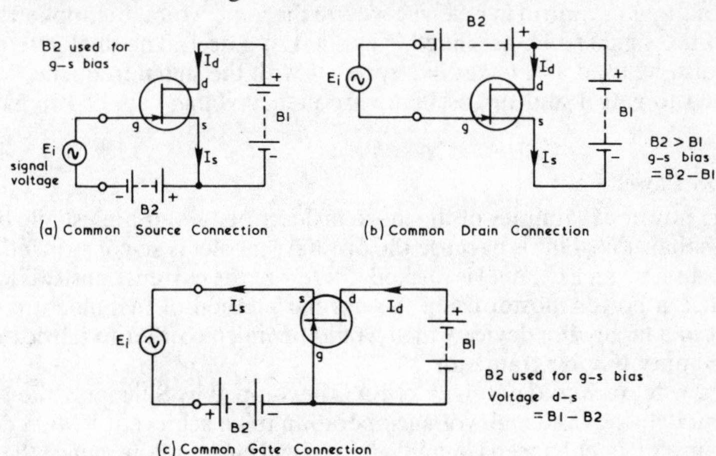

Fig. 4.9 Fet circuit configurations.

Dual-gate Mosfet

Another type of mosfet is one having two independent gates. The basic construction of an n-channel depletion mosfet with dual gates is shown in Fig. 4.10. It consists of three diffused n^+ areas on a P-type substrate with two channels have N-type conductivity formed during manufacture. The current flowing between drain and source may be controlled by a voltage applied to

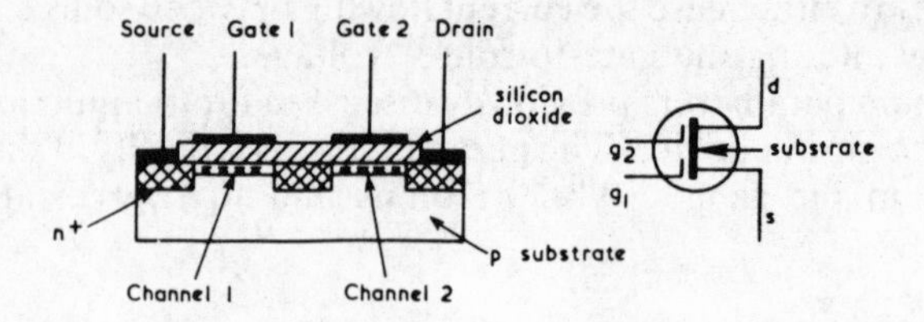

Fig. 4.10 Dual-gate mosfet.

either gate. If gate 1 or gate 2 is made sufficiently negative the current may be cut off.

With an ordinary mosfet the capacitance between drain and gate is about 0·5pF. Although this is a small capacitance it can cause feedback between the output electrode (drain) and input electrode (gate) at high frequencies, resulting in possible oscillation. The introduction of a second gate reduces the capacitance between drain and gate 1 to about 0·02pF provided gate 2 is held at a constant voltage. In effect gate 2 acts as a screen between drain and gate 1.

The gain of the transistor may be altered by varing the steady d.c. voltage on gate 2 with gate 1 used as the signal input electrode. This is useful for automatic gain control in receivers where the a.g.c. voltage is applied to gate 2 and the signal to be amplified is applied to gate 1. The dual-gate mosfet may also be used as a mixer in a receiver with the signal frequency voltage applied to gate 1 and the oscillator frequency voltage applied to gate 2.

VMOS Power Fet

The power capabilities of the mosfets described so far are usually limited to less than 1W. This is because the ordinary mosfet is so constructed that a horizontal surface channel is formed, therefore the current density is low. To produce a power mosfet using the normal method of manufacture would result in a large area device which would be much costlier to fabricate than an ordinary bipolar transistor.

A new fet technology called VMOS (devoloped by Siliconix) allows high current densities and high voltage operation to be achieved. VMOS devices make use of an enhanced channel and vertical current flow, hence the name VMOS (vertical mosfet).

The basic idea of construction for one type is shown in Fig. 4.11. The VMOS fet has a four-layer vertical structure which is formed using a diffusion process. In operating this n-channel enhancement-mode device the gate and drain are both held positive with respect to the source. The voltage applied to the gate induces an n-channel into the p-region on both sides of the V-groove. This allows electrons to flow from the source through the n-channel and N^- area to the N^+ region which is connected to the drain. Current thus flows vertically through the transistor from the source on top to the drain on the bottom of the silicon chip. The higher the gate voltage, the

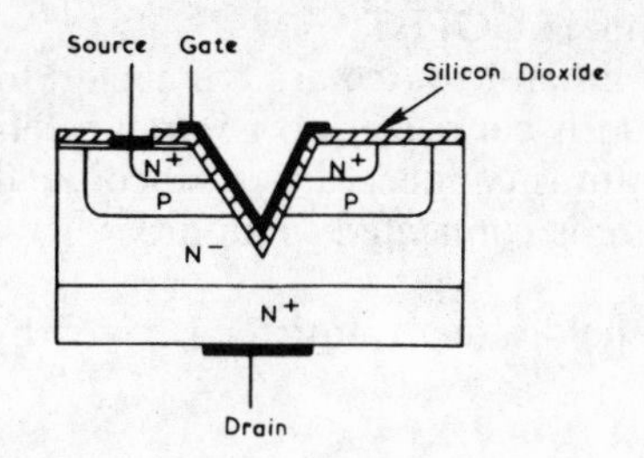

Fig. 4.11 VMOS (enhancement).

wider is the enhanced channel and the greater the current flow. The N^- region (epitaxial layer) is a lightly doped layer and is used to increase the drain-to-source breakdown voltage and to reduce the feedback capacitance.

A typical family of output characteristics for a VMOS power fet is given in Fig. 4.12. These are similar to the ordinary mosfet except for the following:

(a) The drain current is now in amps rather than milliamps;
(b) The output resistance is very high (the curves are practically horizontal);
(c) The spacing between curves is constant above about 0·4A, i.e. the relationship between V_{gs} and I_d is a linear one and not square law.

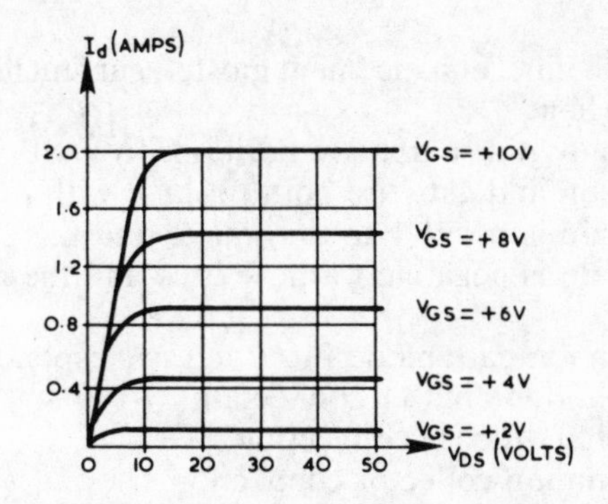

Fig. 4.12 Output characteristic of VMOS power fet.

QUESTIONS ON CHAPTER FOUR

(1) A field-effect transistor is:
  (a) A unipolar device
  (b) A current-operated device
  (c) Very temperature dependent
  (d) A bipolar device.

(2) In an n-channel JUGFET:
 (a) The drain and source are connected to P-type silicon
 (b) The gate is connected to an area of N-type silicon
 (c) The drain is connected to P-silicon and the source to N-silicon
 (d) The gate is connected to an area of P-type silicon.

(3) The g_m of a field-effect transistor is given by:
 (a) $\frac{\delta V_{gs}}{\delta I_g}$
 (b) $\frac{\delta V_{ds}}{\delta I_d}$
 (c) $\frac{\delta I_d}{\delta I_g}$
 (d) $\frac{\delta I_d}{\delta V_{gs}}$.

(4) The value of R_{IN} for a JUGFET is of the order of:
 (a) 1kΩ
 (b) 100Ω
 (c) $10^7\Omega$
 (d) $10^{-6}\Omega$.

(5) In an n-channel enhancement mosfet, current flows between drain and source when:
 (a) Drain and source are both positive with respect to the gate
 (b) Drain and gate are both positive with respect to the source
 (c) Drain is positive to the source and the gate voltage is zero
 (d) Drain is negative to the source and the gate voltage is zero.

(6) The common-gate mode for an fet corresponds to the bipolar:
 (a) Common-emitter connection
 (b) Common-base connection
 (c) Common-collector connection
 (d) Emitter-follower connection.

(7) The type of transistor to be found in the output stage of a 20W audio amplifier would be:
 (a) A JUGFET
 (b) A dual-gate fet
 (c) A VMOS fet
 (d) An n-channel MOSFET.

(Answers on page 180)

CHAPTER FIVE

THE TRANSISTOR AS AN AMPLIFIER

IN VOLUME 1 of this series the main characteristics of a.c. and d.c. amplifiers were discussed. It was noted that all amplifiers have the general property of amplifying, i.e. producing an output signal which is greater than the input signal. All amplifiers are power amplifiers where the output signal power exceeds the input signal power. However, they may be divided into voltage amplifiers, current amplifiers or power amplifiers depending upon whether the main concern is a large voltage, current or power gain. In this chapter we shall consider the basic operation of a small signal voltage amplifier which is intended to serve as an introduction to show how bipolar and unipolar transistors may be used as amplifiers. Further details of other types of amplifiers and their circuits are given in Volume 3.

THE VOLTAGE AMPLIFIER

When dealing with the basic operation of the bipolar transistor it was seen that by altering the base–emitter voltage the collector current could be varied. Similarly with the unipolar transistor the drain current could be altered by varying the gate–source voltage. This property of the two types of transitor is used to achieve amplification by allowing the input signal to vary the bias of the transitor and hence vary the output current. If voltage amplification is required, the output current must be directed through a resistance load (R_L) connected in the output circuit of the transistor, Fig. 5.1.

Selection of Bias Point

When amplifying a signal we ideally require the output signal to be a faithful but magnified replica of the input signal. If the shape of the output signal voltage from the circuits of Fig. 5.1 is to be an exact replica of the input signal voltage, the output signal current flowing in R_L must be a faithful replica of the input signal voltage. To determine how the input current will

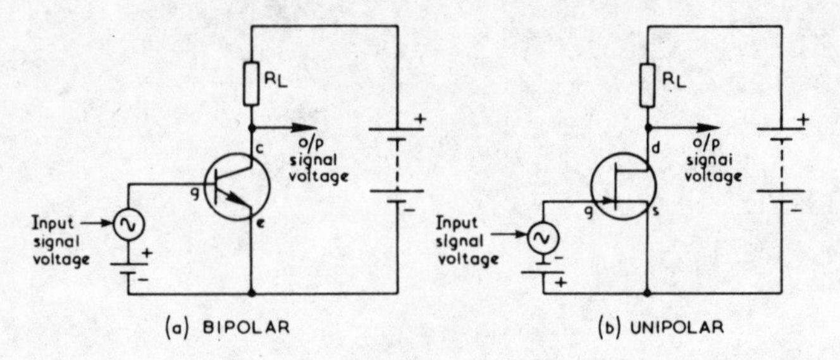

Fig. 5.1 Basic transistor voltage amplifiers.

vary with the input voltage we may use the dynamic mutual characteristic of the transistor. This characteristic shows how the output current varies with changes in the bias voltage when the transistor is used with a particular value of load resistance and supply voltage.

For the output signal current to be identical in shape to the input signal voltage the relationship between the output current and bias voltage should be linear. In practice, there is always some non-linearity which may cause distortion in an amplifier. Thus it is necessary to select a suitable bias point

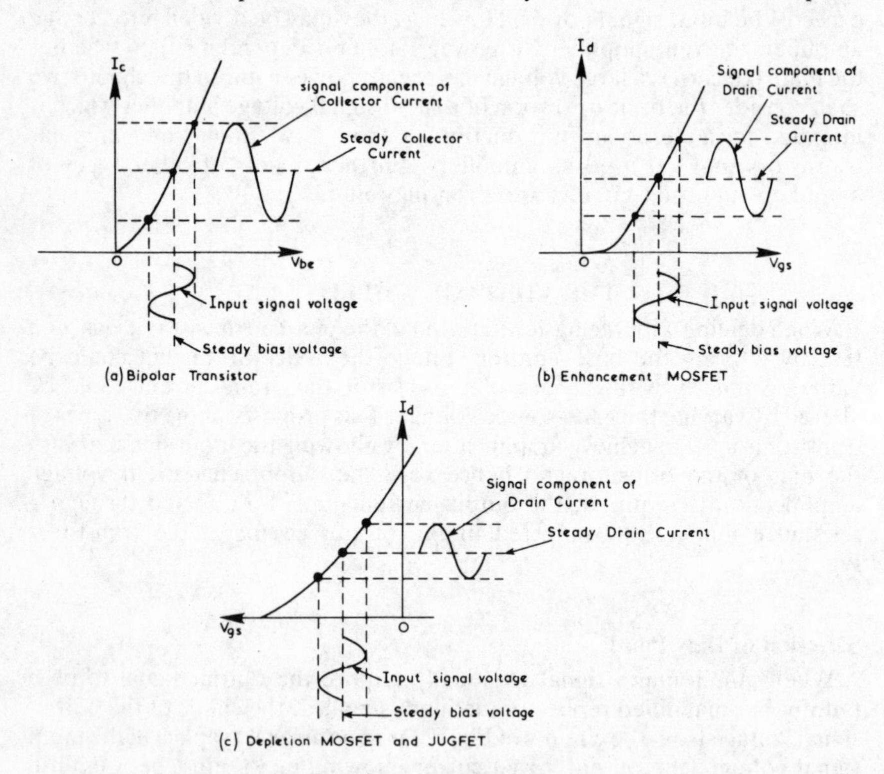

Fig. 5.2 Class-A operation of bipolar and unipolar transistors.

and to limit the amplitude of the input signal voltage so that operation is confined to a portion of the characteristic where the linearity is good. To achieve maximum signal performance, the bias point is chosen so that it lies in the centre of the linear part of the dynamic characteristic as shown in Fig. 5.2. This steady bias voltage will give rise to a steady (or quiescent) output current. When the signal voltage is applied at the input centred on the steady bias voltage, the output current will swing above and below the steady output current level, i.e. the output current consists of a steady or d.c. current with an a.c. component of current superimposed on it. When the output current flows at every instant of a complete input cycle as in Fig. 5.2, the transistor is said to be operated in 'Class-A'. Throughout the remainder of this section Class-A operation will be assumed.

Obtaining the Bias Voltage

We must now discuss practical methods of obtaining bias voltage for the transistor to set the operating point to Class-A.

(a) **Bipolar Transistor Bias.** The simplest method of providing a forward bias voltage or current for a bipolar transistor is to connect a resistor R_B between the supply line V_L and the base if the transistor as shown in Fig. 5.3(a). If the required bias voltage V_{be} is, say, 0·6V and the magnitude of the base current and supply line voltage is known, the value of R_B may be determined as follows.

$$\text{Voltage drop across } R_B = V_L - V_{be}$$

$$\therefore R_B = \frac{V_L - V_{be}}{I_b} \text{ ohms}$$

With a supply voltage of 10V and a base current of 20μA

$$R = \frac{10 - 0{\cdot}6}{20 \times 10^{-6}} \, \Omega = 4{\cdot}7\text{M}\Omega$$

Although apparently satisfactory, this method is rarely used for the following reasons.

(1) An amplifier requires a fixed value of quiescent collector current. With a fixed value of R_B we would obtain a constant value of base current and hence a constant collector current for a transistor with a particular value of h_{FE}. However, in mass-produced transistors the value of h_{FE} is liable to vary over quite a large range of, say, 100 to 500 with transistors of the same type number. Thus with a fixed value of base current, the collector current would vary over a large range of 1:5. This would upset the operating conditions of the amplifier and cause severe distortion.

(2) If the supply line voltage to the amplifier were to vary, the base current of the transistor would also vary since the bias current is approximately proportional to the supply voltage. This would cause a large variation in the

quiescent collector current and would adversely affect the operating conditions of the amplifier.

(3) The arrangement does not provide any protection against variations in ambient temperature. An increase in working temperature of the transistor will increase the leakage current I_{CEO} which will increase the quiescent collector current.

A far better arrangement and one most commonly used is shown in Fig. 5.3(b). First, a resistor $R3$ is placed in the emitter circuit which will thus carry the emitter current. The emitter voltage V_e is given by $I_e \times R3$. A potential divider $R1$, $R2$ is used to feed an approximately constant voltage V_b to the base. The required value of V_b (the voltage across $R2$) is V_e plus the base–emitter voltage drop V_{be}.

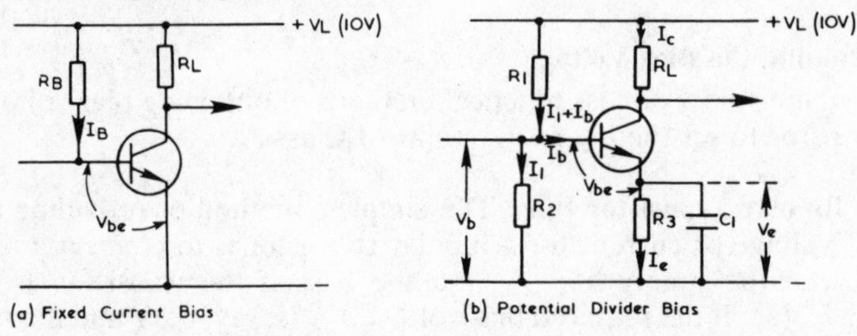

Fig. 5.3 Methods of providing bias for bipolar transistors.

Assume that the value of $R3$ is chosen to give a V_e of 1·4V and that the forward bias V_{be} is 0·6V. The voltage V_b across $R2$ will be 1·4V + 0·6V = 2·0V. In order that the base voltage does not vary appreciably with the base current, the bleeder current I_1 flowing in $R1$ and $R2$ should be large compared with the base current. Suppose I_1 is made $10 \times I_b$. Thus if the base current is 20μA, I_1 will be 200μA which is the current in $R2$. The value of $R2$ is therefore

$$\frac{2}{200 \times 10^{-6}} = 10\text{k}\Omega$$

The current flowing in $R1$ will be $I_1 + I_b$ = 220μA, and the voltage across $R1$ will be $V_L - V_b = 10 - 2 = 8$V. The value of $R1$ is therefore

$$\frac{8}{220 \times 10^{-6}} = 36{\cdot}363\text{k}\Omega\text{, say } 36\text{k}\Omega.$$

In practice the nearest preferred value of 33kΩ or 39kΩ would be used. If a collector current of 2mA is aimed for, then since the emitter current is practically the same the approximate value of $R3$ would be

$$\frac{1{\cdot}4}{2 \times 10^{-3}} = 700\Omega$$

(The nearest preferred value of 680Ω would be used.)

The particular feature of this biasing arrangement will now be explained. Suppose that the emitter current increases. Since this current flows in $R3$, the emitter voltage V_e will rise. Now if I_e increases, I_b must increase but as the bleeder current I_1 is made large compared with I_b the base voltage V_b will remain reasonably constant. In consequence, due to the rise in V_e, the forward bias V_{be} is reduced. This will cause the emitter current to fall thereby practically cancelling the original rise. If the emitter current decreases V_e falls thereby increasing V_{be} (V_b remaining constant) and the emitter current will rise. This rise will oppose the original fall of emitter current.

Thus the action of $R3$ together with the potential divider $R1$, $R2$ supplying constant base voltage V_b, is to stabilise the emitter current and hence the collector current of the transistor. This stabilising action on the collector current will take place if I_c varies as a result of (a) change in h_{FE}; or (b) variation in supply rail voltage; or (c) variation in ambient temperature.

In this description we have been considering only the stabilisation of the d.c. or quiescent conditions of the transistor amplifier. When a signal is applied, the effect of $R3$ (without $C1$) would be to reduce the signal variations of collector current. This is undesirable as the a.c. gain of the amplifier would be reduced. Thus to maintain the a.c. gain, $R3$ is decoupled by the capacitor $C1$ which has such a value that its reactance is very much smaller than the value of $R3$. This is discussed fully in Volume 3 of the series.

An alternative method of biasing which is sometimes used is shown in Fig. 5.4(a). The forward bias current is supplied by $R1$ connected between

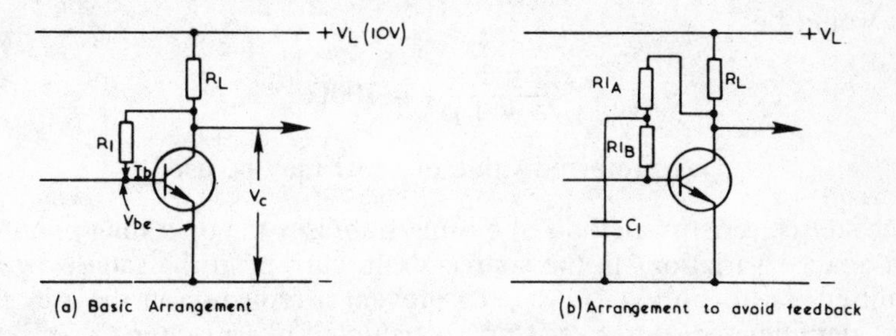

Fig. 5.4 Alternative biasing arrangement for bipolar transistors.

collector and base. Suppose that R_L is 1kΩ and a collector current of 2mA is to be used. The voltage drop across R_L will be $2 \times 10^{-3} \times 10^3 = 2$V. With a supply rail of 10V, the steady collector voltage V_c will be $10 - 2 = 8$V. Now $V_c = V_{R1} + V_{be}$, thus if $V_{be} = 0{\cdot}6$V the voltage drop across $R1$ (V_{R1}) will be $8 - 0{\cdot}6 = 7{\cdot}4$V. If the base current is 20μA as in the previous example, the value of $R1$ will be

$$R1 = \frac{7{\cdot}4}{20 \times 10^{-6}} = 370000\Omega$$

(The nearest preferred value of 390kΩ would be used.)

This method also provides a stabilising action on collector current. If, say, I_c increases for any of the reasons discussed there will be an increase in the volt drop across R_L and the collector–emitter voltage V_c will fall. In consequence the base–emitter voltage V_{be} and I_b will fall thereby reducing the collector current until equilibrium is reached. Since $R1$ is connected between collector and base any signal voltage variations across R_L will also be fed back to the base and reduce the a.c. gain of the amplifier. To avoid a reduction in the a.c. gain the circuit may be modified to that shown in Fig. 5.4(b). $R1_A + R1_B$ is made the same value as previously calculated and the operation as regards d.c. is identical. If, however, $C1$ is included and given a value so that its reactance is small compared with $R1_A$ value, the a.c. feedback will be eliminated.

(b) **Unipolar Transistor Bias.** An n-channel JUGFET is normally operated with its gate negative with respect to its source and a common method of achieving this is shown in Fig. 5.5(a). Here a resistor R_s is connected between the source and common line such that it carries the source current I_s which will provide a voltage drop (V_s) across R_s with polarity as shown. This voltage drop provides the required value of bias voltage. The gate is connected to the common line via a resistor R_G and since the gate current is minute there will be negligible voltage drop across R_G if it is not made more than 1–2MΩ. Thus the gate potential is the same as that of the common line, i.e. the same as the potential at the lower end of R_s. Therefore the gate is negative to the source by the voltage drop across R_s.

With a source current of, say, 4mA and a required bias of 2·8V the value of R_s would be

$$\frac{2{\cdot}8}{4 \times 10^{-3}} = 700\Omega$$

(A preferred value of 680Ω may be used.)

The source resistor method of biasing stabilises the operating point of the fet against variations in the source–drain current in the same way as the emitter resistor in Fig. 5.3(b). To prevent a reduction in the a.c. gain of the amplifier the source resistor is decoupled by capacitor C_1.

The biasing arrangement of Fig. 5.5(b) is sometimes used as this is more effective in dealing with wide spreads in the g_m of JUGFETS of the same type. A potential divider $R1$, $R3$ provides a voltage on the gate (V_g) which is positive to the common line. This voltage is made less than the voltage drop (V_s) across the source resistor R_s and the difference between V_s and V_g gives the required bias voltage. For example if V_s is 5V and V_g is 2V the effective bias voltage is 3V with the gate negative to the source. This method allows a larger value of R_s to be used for a given source-to-drain current and gives better stabilisation of the drain current with spreads in g_m. The presence of the potential divider $R1$, $R3$ would tend to lower the input resistance of the amplifier if the junction of these resistors were connected directly to the gate. As regards a.c., $R1$ and $R3$ are effectively in parallel with each other

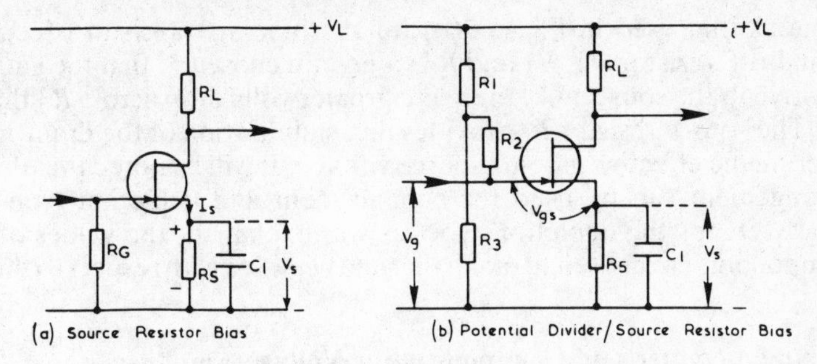

Fig. 5.5 Methods of providing bias for jugfets.

and would shunt the input terminals of the fet. The use of $R2$ effectively increases the input impedance to

$$R2 + \frac{R1 \times R3}{R1 + R3}.$$

With an n-channel depletion mosfet the gate is biased negatively with respect to the source and either of the arrangements shown in Fig. 5.6. may be used. These are directly equivalent to the methods used in Fig. 5.5. for

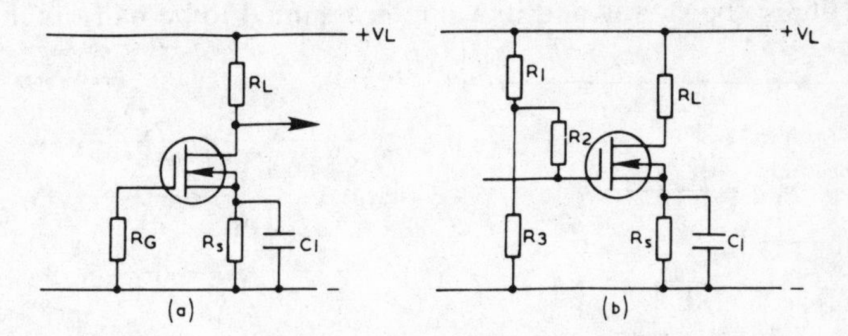

Fig. 5.6 Biasing arrangements for n-channel depletion mosfets.

jugfets, the principle of biasing and stabilisation being the same. When an n-channel enhancement mosfet is used the gate must be biased positively to the source and the circuit of Fig. 5.7 may be used. This arrangement is the

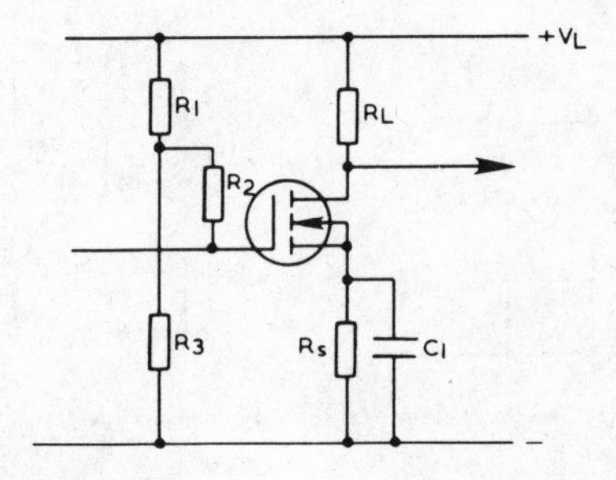

Fig. 5.7 One method of biasing an n-channel enhancement mosfet.

same as that used in Fig. 5.3(b) for the bipolar transistor except for the 'stand-off' resistor $R2$. $R1$ and $R3$ values are chosen so that the gate is made positive to the source, i.e. there is a greater volts drop across $R3$ than across R_s. The source resistor R_s provides d.c. stabilisation of the drain current by altering the effective gate-to-source voltage. It will be noted that this biasing arrangement can be used for enhancement and depletion type mosfets. However, for the depletion type, as was explained, the values of the bias components are chosen to make the gate negative with respect to the source.

Common-emitter and Common-source Voltage Amplifier

The common-emitter and common-source modes of operation are the connections most frequently used. Basic operation of the common-emitter voltage amplifier is shown in Fig. 5.8(a). Forward bias for the transistor is provided by $R1$, $R2$ with $R3$ used for d.c. stabilisation of the collector current. $C1$ prevents the d.c. base bias voltage being shorted out via the input signal source. This capacitor will have a low reactance to the input signal, assumed to be sinusoidal.

The steady forward bias V_{be} applied to the transistor will give rise to a steady value of collector current and a steady voltage drop across the collector load R_L. Thus the voltage between the collector and negative line (V_c) will also be steady and its value is assumed to be half the line supply

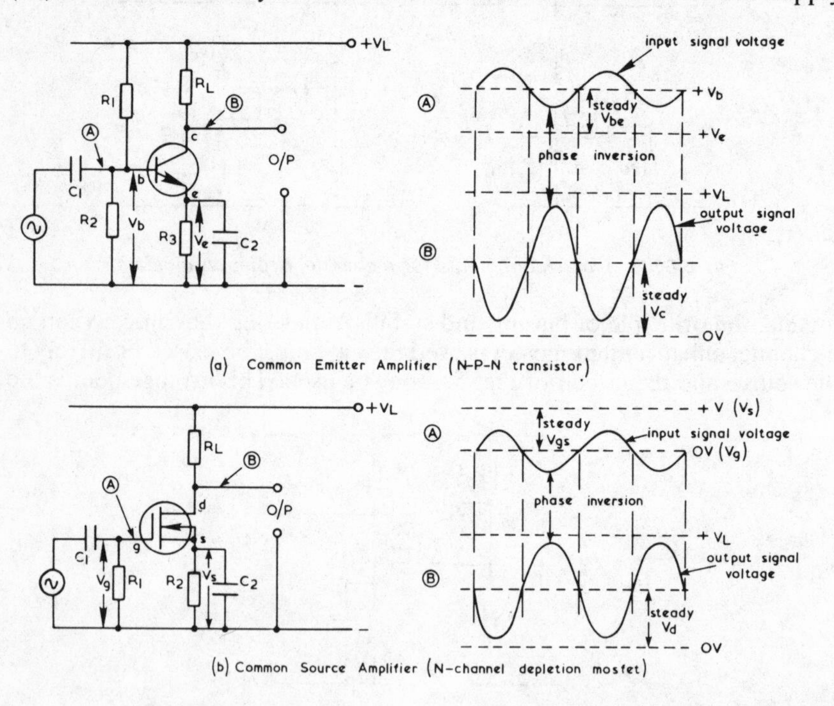

Fig. 5.8 Common-emitter and common-source amplifiers.

voltage, see waveform B. When the input signal is applied through $C1$ it causes the forward bias to alter. On positive half-cycles, the forward bias will be increased thereby increasing the collector current and the voltage developed across R_L. In consequence, V_c will fall below its steady level. During negative half-cycles of the input, the forward bias will be reduced thereby decreasing the collector current and the voltage developed across R_L. As a result V_c will rise above its steady level. In this way a magnified but **inverted** version of the input signal voltage is obtained at the output. The output signal voltage may be taken from across R_L or between the collector and the negative line since, as regards a.c., the positive and negative rails are short circuited.

The operation of a common source amplifier is similar. With an n-channel depletion mosfet the gate is biased negatively with respect to the source and in Fig. 5.8(b) is obtained by the voltage drop across $R2$. The input signal is applied to the gate via the low reactance of $C1$. This capacitor will block any d.c. component present in the signal. The steady bias voltage V_{gs} of waveform A will give rise to a steady drain current thereby producing a steady voltage drop across the load resistor R_L. Thus the voltage between drain and the common negative line (V_d) will also be steady and is assumed to be half the line supply voltage, see waveform B. On positive half-cycles of the input signal the bias will be reduced causing an increase in the drain current and the voltage drop across R_L, and V_d will fall below its steady level. Conversely, on negative half-cycles of the input the bias is increased causing a decrease in drain current and the voltage drop across R_L. As a result V_d will rise above its steady level. Again, a magnified but **inverted** version of the input signal voltage is obtained at the output. The voltage gain of the common-source amplifier is less than that of the common-emitter amplifier.

Common-base and Common-gate Voltage Amplifier

Circuit arrangements for the common-base and common-gate amplifiers are shown in Fig. 5.9. With the common-base amplifier of Fig. 5.9(a), $R1$ and $R2$ bias the base positively to the emitter in the same way as for the common-emitter amplifier. $R3$ is the emitter stabilising resistor, decoupled by $C2$ to prevent a.c. negative feedback. $C1$ is added to connect the base to the common line as regards a.c. This is essential to prevent the a.c. component of base current flowing in $R2$, i.e. it makes the circuit common-base by connecting the base to the common line.

The input signal is now applied to the emitter and the signal source must provide a path for the d.c. component of emitter current. Waveforms A and B show the basic operation for sinewave input. As for common-emitter operation, the steady V_{be} gives rise to a steady I_c and this produces a steady voltage drop across R_L. In consequence the collector-to-common line voltage V_c is also steady. When the input signal is on a positive half-cycle at the emitter the forward bias is reduced causing I_c to fall and V_c to rise above its steady level. Conversely, on negative half-cycles of the input signal the

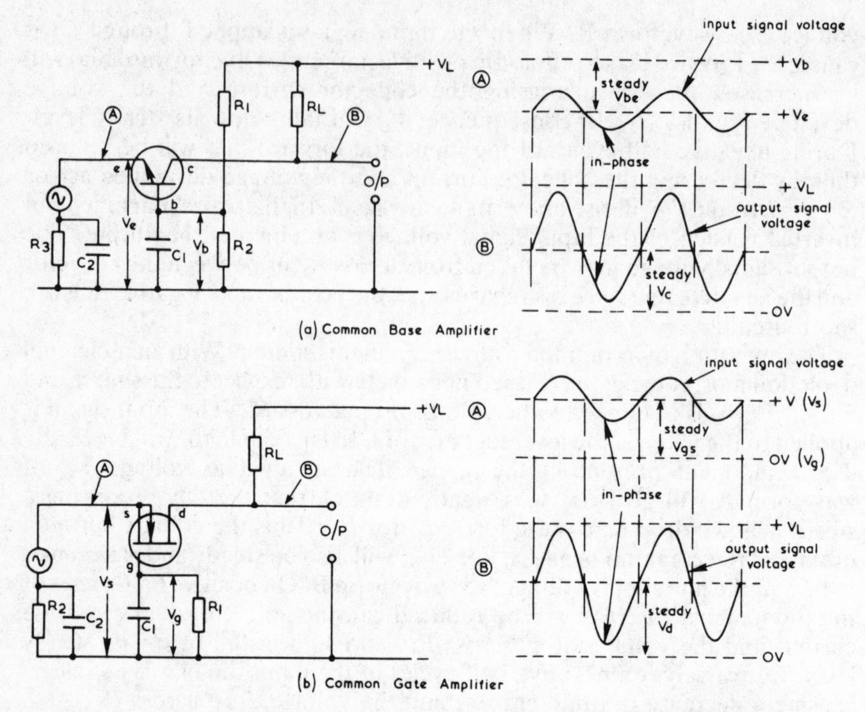

Fig. 5.9 Common-base and common-gate amplifiers.

forward bias is increased causing I_c to increase and V_c to fall below its steady level. Thus at the output a magnified replica of the input signal is obtained which is in phase with the input signal.

In the common-gate amplifier of Fig. 5.9(b), *R*2 provides the required bias and is decoupled by *C*2 to prevent a.c. feedback. Resistor *R*1 provides a d.c. return to the common line for the gate and *C*1 connects the gate to the common line as regards a.c. The input signal is applied to the source electrode and the signal source must provide a d.c. path for the source current.

Positive half-cycles of the input signal increase the bias thereby reducing the drain current and causing the drain-to-common line voltage to rise above its steady level. Conversely, on negative half-cycles the bias is reduced, the drain current is increased and the output voltage falls below its steady level. An amplified version of the input signal is obtained at the output and, as for the common-base circuit, input and output signal voltages are in phase. The voltage gain of the common-gate amplifier is less than that of a common-base amplifier.

Common-collector and Common-drain Amplifiers

In the common-collector amplifier of Fig. 5.10(a), the potential divider

$R1$, $R2$ forward bias the transistor in the usual way. The output signal voltage is developed across R_L placed in the emitter circuit and this resistor also provides d.c. stabilisation of the operating point.

The steady forward bias V_{be} will cause a steady emitter current to flow and produce a steady voltage drop V_e across the emitter resistor R_L. When the input signal voltage goes positive the forward bias is increased causing an increase in emitter current and in the voltage drop across R_L. During negative half-cycles of the input the forward bias is reduced causing the emitter current to decrease and less voltage to be developed across R_L. Input and output signal voltages are thus in phase in the common-collector connection but there is however **no voltage gain**. This is because the signal voltage across R_L provides a.c. feedback reducing the effective signal voltage applied between base and emitter. If the output signal voltage were able to rise to an amplitude such that it equalled the input signal voltage (which is impossible) the net signal voltage between base and emitter would be zero and so the output would be zero. Thus the voltage gain is less than unity, permitting a net signal voltage between base and emitter to drive the transistor. Because of the a.c. feedback the input resistance of the common-collector amplifier is higher than the common-emitter amplifier, lying typically in the range 5–500kΩ. The output resistance is low and lies in

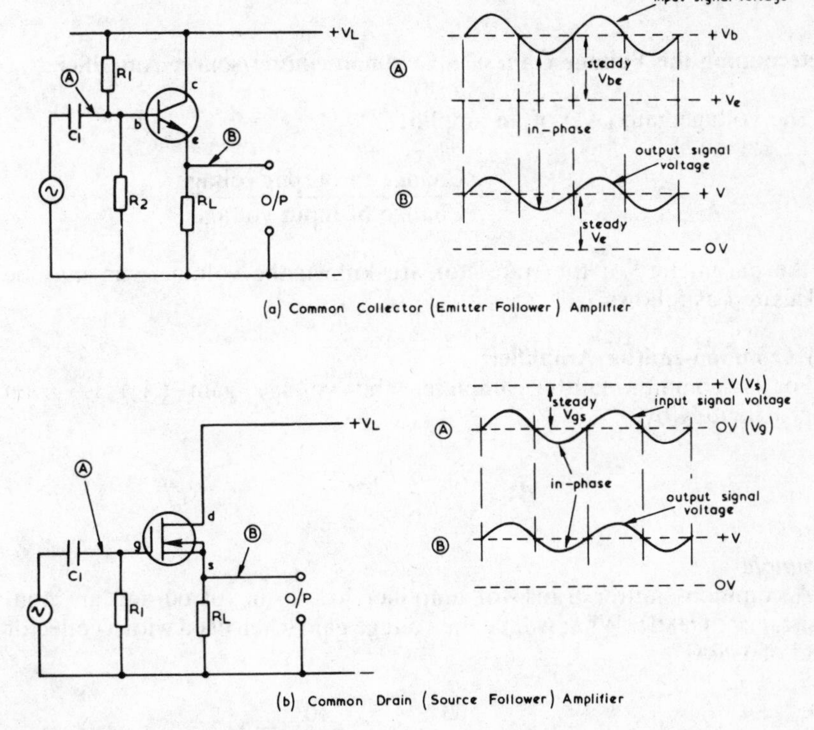

Fig. 5.10 Common-collector and common-drain amplifiers.

the range 50–500Ω. The main use of the common-collector or 'emitter-follower' amplifier is as a buffer amplifier connected between a high-impedance source and a low-impedance load.

The equivalent circuit arrangement for the unipolar transistor is given in Fig. 5.10(b). Here the steady voltage drop across R_L biases the gate negative with respect to the source in the normal way. The signal voltage output across R_L has an in-phase relationship with the input signal voltage as for the common-collector circuit and a voltage gain less than unity.

The table below summarises the features discussed for the three modes of amplifier connection:

Connection	**Voltage Gain**	**Phase relationship between input and output signal voltages**
Common Emitter	High	Phase reversal
Common Base	High	In phase
Common Collector	<1	In phase
Common Source	Medium	Phase reversal
Common Gate	Medium	In phase
Common Drain	<1	In phase

Determining the Voltage Gain of a Common-emitter/source Amplifier

The voltage gain (A_v) of an amplifier

$$= \frac{\text{Change of output voltage}}{\text{Change of input voltage}}.$$

If the parameters of the transistor are known the voltage gain may be calculated as follows:

(a) **Common-emitter Amplifier**

For a common-emitter amplifier the voltage gain (A_v) is given approximately by

$$A_v = \frac{h_{fe} \times R_L}{R_{IN}} \text{ (or } A_v = g_m R_L)$$

Example

A common-emitter transistor amplifier has an h_{fe} of 60 and an input resistance of 800Ω. What will be the voltage gain when used with a collector load of 1·5kΩ?

$$A_v = \frac{60 \times 1{\cdot}5 \times 10^3}{800} = 112{\cdot}5.$$

(b) **Common-source Amplifier**

For a common-source amplifier the voltage gain (A_v) is given by

$$A_v = \frac{g_m \times R_L \times r_{ds}}{r_{ds} + R_L}$$

where r_{ds} is the drain-to-source a.c. resistance.

If r_{ds} is much greater than R_L which is often the case, the expression may be simplified to

$$A_v \approx g_m \times R_L.$$

Example

An fet common-source amplifier has a g_m of 5×10^{-3}S and an r_{ds} of 100kΩ. What will be the voltage gain when used with a drain load of 5kΩ?

Since r_{ds} is much greater than R_L the simplified expression may be used, thus

$$\begin{aligned} A_v &\approx g_m \times R_L \\ &\approx 5 \times 10^{-3} \times 5 \times 10^3 \\ &\approx 25 \end{aligned}$$

The voltage gain for an fet amplifier is appreciably smaller than for a bipolar transistor amplifier since the g_m is much lower (3mS for an fet as opposed to 40mS for a bipolar device).

Use of a Load Line

The voltage and current gain of a bipolar transistor may also be determined with the aid of a load line drawn on the output characteristics.

Consider Fig. 5.11 which shows the steady voltages and current that exist in a common-emitter amplifier. Under all operating conditions the supply voltage V_L must be equal to the sum of the voltages between the collector

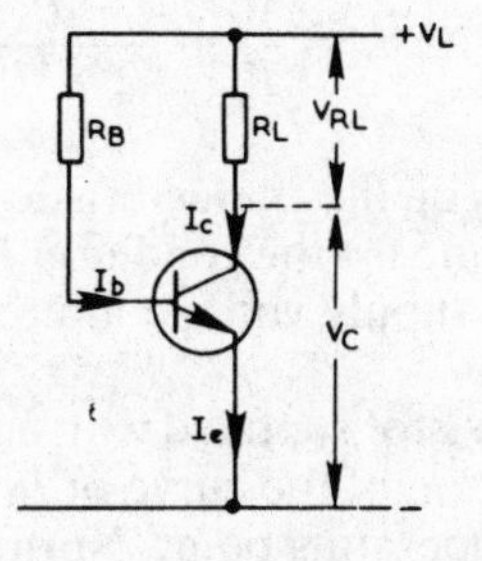

Fig. 5.11 Common-emitter amplifier.

and common line (V_c) and the voltage drop V_{R_L} across the collector load resistor, i.e.

$$V_L = V_{R_L} + V_c$$

$$\text{Now } V_{R_L} = I_c R_L$$

$$\text{Therefore } V_L = I_c R_L + V_c \qquad (1)$$

Equation (1) is of the form $y = mx + c$ and is the equation to a straight line. To draw the line it is only necessary to plot two points and connect them.

Point 1

$$\text{Let } I_c = 0$$

$$\therefore \text{ using equation (1),}$$

$$V_L = V_c.$$

Point 2

$$\text{Let } V_c = 0$$

$$\therefore \text{ using equation (1),}$$

$$V_L = I_c R_L$$

$$\text{or } I_c = \frac{V_L}{R_L}$$

Thus if the two points $I_c = 0$, $V_c = V_L$ and $V_c = 0$, $I_c = V_L/R_L$ are plotted on the output characteristics and the points joined, a load line is formed for a particular value of load resistance and supply voltage.

It will be assumed that the output characteristics of Fig. 5.12 are those of the transistor used in the circuit of Fig. 5.11 and that the supply V_L is 10V and R_L is 2kΩ.

Point 1 is given by $I_c = 0$, $V_c = V_L = 10\text{V}$.

Point 2 is given by $V_c = 0$, $I_c = \frac{V_L}{R_L} = \frac{10}{2 \times 10^3} = 5\text{mA}$.

These two points are shown on the characteristics with a line drawn between them. This line is the load line for the condition $R_L = 2\text{k}\Omega$ and $V_L = 10\text{V}$. If the load is changed or the supply voltage altered a new load line must be constructed.

With the base of the transistor supplied with a steady base current of, say, 30μA via R_B, the interception of the curve of $I_b = 30\mu\text{A}$ with the load line (marked X) gives the d.c. operating point. Normally, this point is chosen to lie towards the middle of the load line. By projecting down and across from

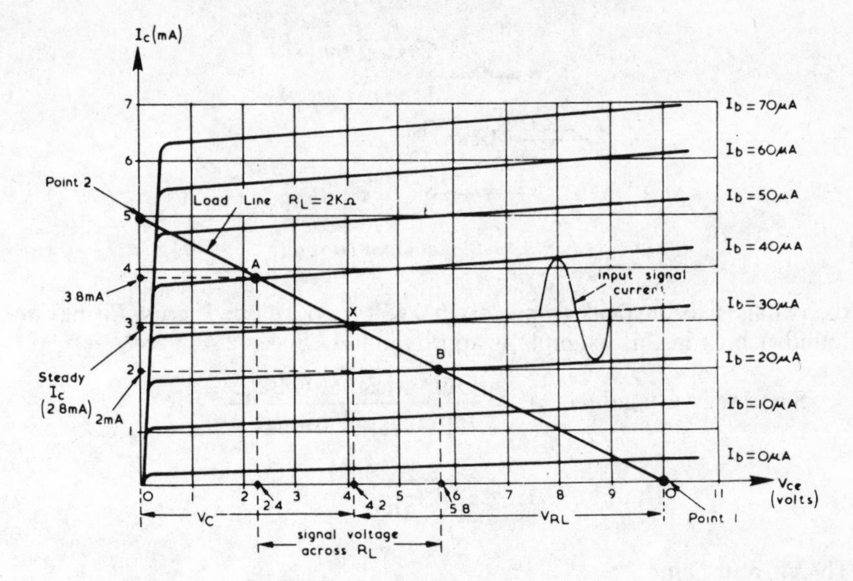

Fig. 5.12 Common-emitter output characteristics and load line.

point X the steady V_c and I_c may be found, which in this example are approximately 4·2V and 2·8mA respectively.

The product of the supply voltage V_L and the steady I_c gives the d.c. power taken from the supply.

$$\text{Therefore d.c. power taken from the supply} = 10 \times 2{\cdot}8 \times 10^{-3}\text{W} = 28\text{mW}.$$

The d.c. power dissipated in the load R_L is the product of the voltage across R_L and the steady collector current.

$$\text{Thus d.c. power dissipated in } R_L = (10 - 4{\cdot}2) \times 2{\cdot}8 \times 10^{-3}\text{W} = 16{\cdot}24\text{mW}.$$

The d.c. power dissipated at the collector of the transistor is the product of the steady V_c and the steady I_c.

$$\text{Thus d.c. power dissipated at collector} = 4{\cdot}2 \times 2{\cdot}8 \times 10^{-3}\text{W} = 11{\cdot}76\text{mW}.$$

Note that the sum of the d.c. powers dissipated at the collector and in R_L is equal to the d.c. power taken from the supply.

If the input signal voltage v_i and the input resistance R_{IN} are known the voltage gain may be determined as follows. Suppose that v_i is 20mV peak-to-peak and R_{IN} is 1kΩ, then the input current (base current) i_i is v_i/R_{IN} = 20μA peak-to-peak, see Fig. 5.13. Thus the operating point on the load line moves between points A and B, i.e. 10μA peak swing either side of the steady base current of 30μA. By projecting down from points A and B to

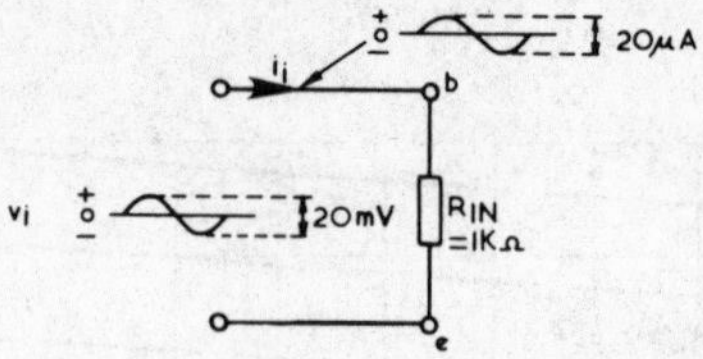

Fig. 5.13 Signal input current.

the voltage axis the peak-to-peak signal voltage developed across R_L may be found, which in this example is approximately 5·8 − 2·4 = 3·4V.

$$\text{Since the voltage gain } A_v = \frac{\text{Output signal voltage}}{\text{Input signal voltage}}$$

$$\text{Therefore } A_v = \frac{3{\cdot}4}{20 \times 10^{-3}} = 170.$$

The current gain

$$A_i = \frac{\text{Output signal current}}{\text{Input signal current}}$$

and may be obtained directly using the load line. The peak-to-peak output current may be found by projecting across to the current axis from points A and B which in this example is 3·8 − 2 = 1·8mA. The peak-to-peak input current is 20μA.

$$\text{Thus } A_i = \frac{1{\cdot}8 \times 10^{-3}}{20 \times 10^{-6}} = 90.$$

Additional information that may be obtained from the load line includes an estimation of the amount of distortion that may be present in the output. This may be found by noting whether the increments AX and BX are equal. If these distances are the same then the peak swings either side of the steady operating point will be equal and distortionless output will be obtained. In our example the distance AX is slightly greater than the distance BX, thus there will be some distortion in the output.

Bottoming and Clipping

When a large input signal is applied to a transistor the output may be severely flattened on one or both peaks depending upon the operating conditions. Consider the load line CD of Fig. 5.14. where a steady base current of 40μA results in a quiescent operating point X. If the peak base input signal current is 40μA the operating point will move up the load line to

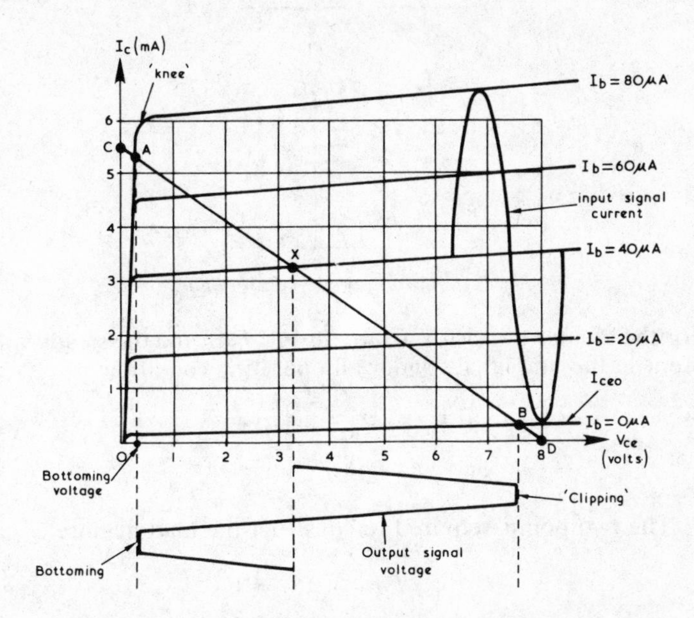

Fig. 5.14 Clipping and bottoming.

point A on one half-cycle and down the load line to point B on the other half-cycle.

At point A where the load line cuts the knee of the characteristic for $I_b = 80\mu A$, the collector-to-emitter voltage falls to an extremely small value (a fraction of a volt). If the base current is increased above $80\mu A$ there will be no further reduction in the collector-to-emitter voltage which remains at its minimum level. This condition is known as 'bottoming' and results in a flattening of the negative peak of the output signal voltage.

Similarly, at point B where the collector-to-emitter voltage is practically the same as that of the supply, the positive peak of the output voltage will be flattened when the base current reaches zero or the signal input attempts to reverse bias the transistor. This condition is called 'clipping' and is the supply rail limit in rise of the output voltage. The clipping level of the output signal voltage is slightly less than the supply voltage due to the voltage drop across R_L produced by the leakage current I_{ceo}.

If the quiescent operating point is chosen for maximum output then clipping and bottoming will occur approximately at the same point on both half-cycles. If the operating point is moved up the load line, bottoming will occur before the clipping level is reached or vice versa with point X lower down the load line.

The load line for a field-effect transistor may be constructed in the same

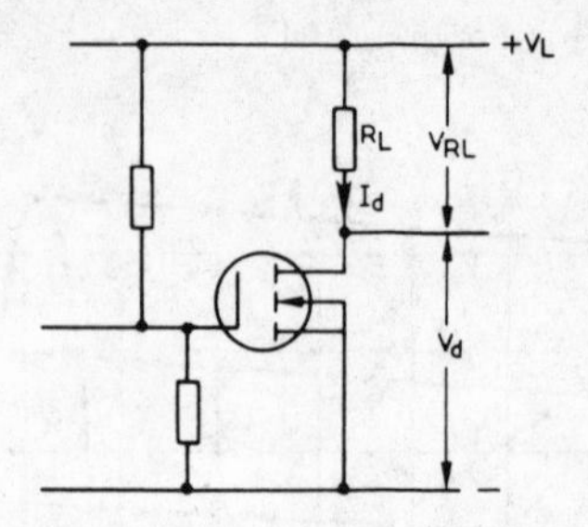

Fig. 5.15 Common-source amplifier.

way as for the bipolar transistor. Consider Fig. 5.15 and the steady voltages and current in the amplifier. Under all operating conditions

$$V_L = V_{RL} + V_d$$
$$\text{or } V_L = I_d R_L + V_d.$$

The two points required to construct the load line are:

$$\text{(i) } I_d = 0,\ V_L = V_d$$
$$\text{and (ii) } V_d = 0,\ I_d = \frac{V_L}{R_L}$$

A load line for $R_L = 2\text{k}\Omega$ and $V_L = 20\text{V}$ is given in Fig. 5.16 superimposed

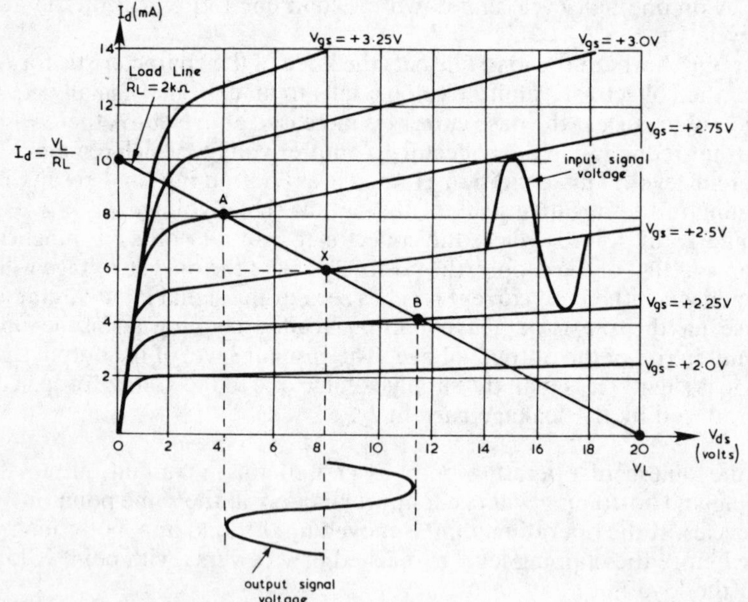

Fig. 5.16 Common-source output characteristics and load line.

on the output characteristics. With a steady bias voltage of +2·5V for an n-channel enhancement mosfet, the steady operating point is at X providing a quiescent V_{ds} of 8V and a quiescent I_d of 6mA. Assuming an input signal voltage of 0·5V peak-to-peak, the operation moves between the limit points of A and B on the load line. The output signal voltage varies between the limits of approximately 4V and 11·5V, a peak-to-peak variation of about 7·5V.

$$\text{The voltage gain } A_v \text{ is thus } \frac{7 \cdot 5}{0.5} = 15.$$

QUESTIONS ON CHAPTER FIVE

(1) With reference to Fig. 5.3(b) on page 64, the purpose of $R3$ is to:
(a) Forward bias the transistor
(b) Provide a.c. feedback
(c) Reverse bias the transistor
(d) Stabilise the d.c. working point.

(2) The magnitude of the current I_1 in Fig. 5.3(b) is usually made about:
(a) $10 \times I_e$
(b) $I_b/10$
(c) $10 \times I_b$
(d) $10 \times I_c$.

(3) The advantage of the biasing arrangement of Fig. 5.5(b) on page 67 over that of Fig. 5.5(a) is that it:
(a) Provides a smaller bias
(b) Provides a larger bias
(c) Deals better with spreads in g_m
(d) Has a higher input resistance.

(4) In a resistive-loaded, common-base amplifier:
(a) There is no phase reversal between input and output signal voltages
(b) The output resistance is low and the input resistance high
(c) The output current is the base current
(d) The input current is the base current.

(5) The voltage gain of a common-drain amplifier would be typically:
(a) 100
(b) 25
(c) 10
(d) 0·99.

(6) The steady voltage between the collector and common line of a common-emitter amplifier is 5V and the steady collector current is 2mA. If the supply rail voltage is 13V the value of the collector load resistance is:
(a) 4kΩ
(b) 2kΩ
(c) 16kΩ
(d) 9kΩ

(7) The voltage gain A_v of a common-emitter amplifier is given approximately by:
(a) $Av = \frac{R_L}{h_{fe}}$
(b) $Av = \frac{h_{fe} \times R_L}{R_{IN}}$
(c) $Av = \frac{R_L}{g_m}$
(d) $Av = \frac{R_L}{h_{fe}}$.

(8) The voltage gain A_v of a common-source amplifier is given approximately by:
(a) $A_v = g_m \times R_L$
(b) $A_v = g_m \times R_{IN}$
(c) $Av = \frac{g_m}{R_L}$
(d) $Av = \frac{g_m}{R_{IN}}$.

(9) The two extreme points on a load line drawn on the output characteristics of a common amplifier are given by Point 1 $I_c = 0$, $V_c =$ 10V and Point 2 $V_c = 0$, $I_c = 2{\cdot}0$mA. The values of the supply rail voltage and the collector load resistance are respectively:
(a) 15V and 1kΩ
(b) 10V and 5kΩ
(c) 5V and 20kΩ
(d) 10V and 20kΩ.

(10) If the supply rail voltage in Fig. 5.16 is reduced from 20V to 16V and a load line is drawn for this condition, it will:
(a) Have a greater slope than the original load line
(b) Lie parallel to the original load line
(c) Coincide exactly with the original load line
(d) Have less slope than the original load line.

(Answers on page 180)

CHAPTER SIX

OTHER SEMICONDUCTOR DEVICES

THYRISTORS

The thyristor family embraces any semiconductor switch whose 'on'/'off' action depends upon a regenerative effect in a p-n-p-n sandwich. These devices can be made to conduct easily in one direction only or in both directions. They are available with two, three or four terminals.

(a) Silicon Controlled Rectifier (SCR)

The most commonly used member of the thyristor family is the SCR or 'reverse blocking triode thyristor'. It has three terminals and is unidirectional. Manufacturing methods include planar and alloy-diffused techniques. Figure 6.1 illustrates the alloy-diffused method. The centre n-type silicon wafer is diffused on both faces with a p-type impurity thus forming a p-n-p structure. A connection to one of the p-regions forms the anode electrode. A pellet of n-type impurity is then alloyed to the other p-region to form the **cathode** electrode. The **gate** terminal is bonded to the lower p-region which constitutes the control layer. This constitutes what is called a 'cathode controlled SCR'. The gate terminal may be brought out from the upper n-region producing an 'anode controlled SCR'. Circuit symbols for both types are given in Fig. 6.2.

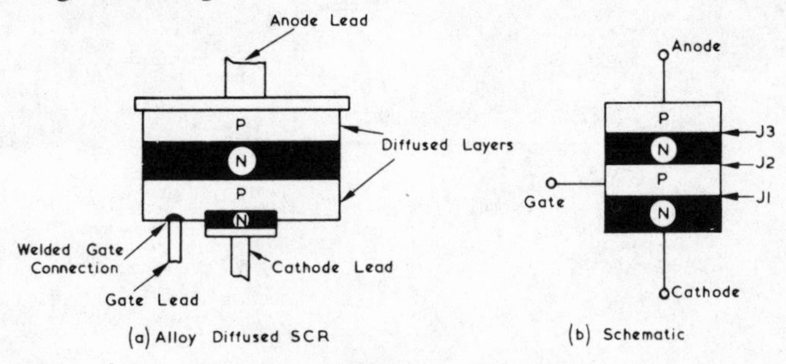

Fig. 6.1 Silicon controlled rectifier.

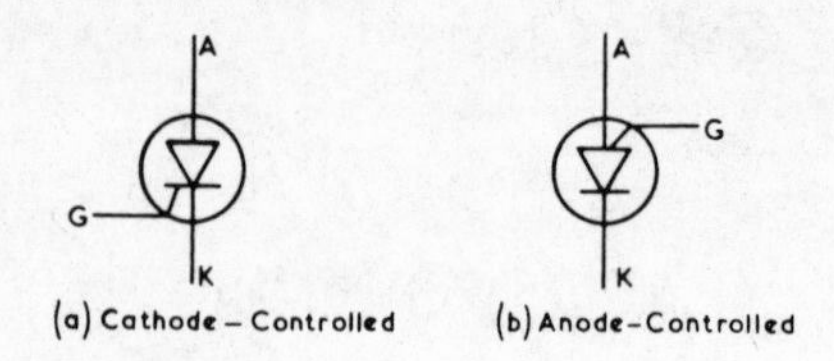

Fig. 6.2 SCR circuit symbols.

Between the anode and cathode leads of the device there are thus four layers with 'p' and 'n' material alternately arranged as shown in schematic form in Fig. 6.1(b). At first sight the structure appears to form three separate diode junctions (*J*1, *J*2 and *J*3). Let us first examine this '3-diode' concept and see if we can use it to explain the action of the device. Suppose the anode is made negative to the cathode but no connection is made to the gate. *J*1 and *J*3 would then be reverse biased but *J*2 forward biased. Thus very little current would flow from cathode to anode until the applied voltage was sufficient to break down the junctions *J*1 and *J*3. This is true of the device in that it behaves as a reverse biased diode when the anode is negative to the cathode (see the reverse blocking region of Fig. 6.3). If the supply voltage polarity is now reversed making the anode positive to the cathode, *J*1 and *J*3 will be forward biased and *J*2 reverse biased. Again the current would be small until the applied voltage was high enough to break down *J*2. This time however when breakdown occurs the characteristic is not like that of a diode. As Fig. 6.3 shows the voltage drop across the device rapidly falls to a very low value. Thus the '3-diode' concept falls down at this point because the three diodes are not isolated from one another.

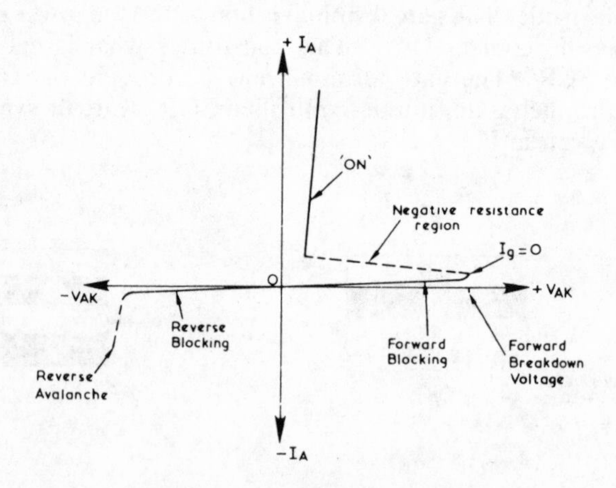

Fig. 6.3 Voltage–current characteristics of SCR.

The p-n-p-n structure is best visualised as consisting of two transistors one of which is a p-n-p and the other an n-p-n, directly coupled as shown in Fig. 6.4(c). The internal feedback loop formed allows the collector current of one transistor to feed the base of the other. Consider now the operation when the anode is made positive to the cathode but the magnitude of the voltage is insufficient to cause breakdown, i.e. the device is in the forward blocking region of Fig. 6.3 and the collector–base junction of both transistors is reverse biased. Now the collector–base leakage current of, say, TR1 will be amplified by TR1 and the resulting collector current fed to TR2 base. This input current will be amplified by TR2, the collector current of which will be fed back to TR1 base for subsequent amplification. The magnitude of the current I flowing between anode and cathode depends upon the current gain around the internal feedback loop. The loop gain is the product of the individual current gains of the two transistors. Provided the loop gain is less than one, the current I is small. If the loop gain is equal to one, the loop becomes regenerative and each transistor drives the other into saturation. At this point all junctions assume a forward bias and the current I rapidly rises to a large value, being limited only by the external circuit resistance.

During manufacturing it is arranged that the loop gain is less than unity at low currents and voltages. To turn the SCR 'on' it is necessary to increase the loop gain. It is a property of transistors that the gain increases with increase in emitter current. There are several mechanisms that will initiate a rise in emitter current and hence a rise in current gain. The important mechanisms are: (a) increasing the collector–emitter voltage; (b) forward biasing the base–emitter junction, i.e. providing base current; and (c) allowing light energy to fall on the centre junction.

Mechanism (a) is that responsible for turning 'on' the SCR with zero gate current (Fig. 6.3). It occurs when the collector–emitter voltage is raised to such a level that the energy of the leakage current carriers is sufficient to

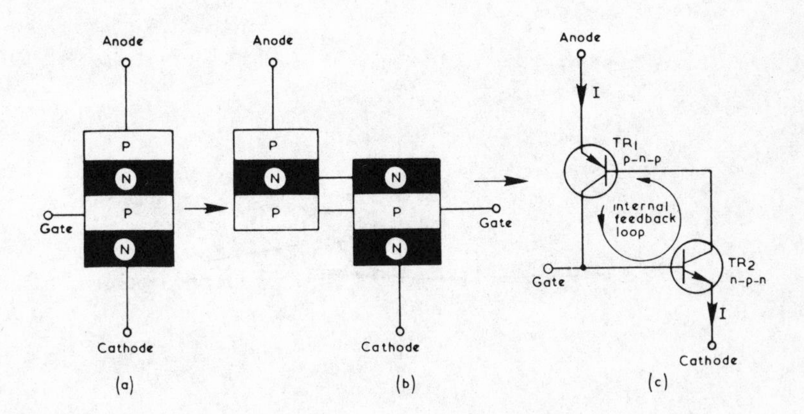

Fig. 6.4 Two-transistor analogy of SCR.

cause the removal of additional carriers from the collector–base region. These carriers in their turn dislodge more carriers which add to the numbers arriving at the collector. This results in an avalanche breakdown and the current increases sharply. As a result the gain of both transistors is increased and when the loop gain reaches unity the SCR comes 'on'.

The usual way of switching 'on' an SCR is via mechanism (b) by forward biasing the base–emitter junction of one of the analogous transistors. With a cathode-controlled SCR the gate is made positive to the cathode and will thus forward bias the base–emitter junction of the analogous n-p-n transistor TR2 of Fig. 6.4(a). For an anode-controlled SCR, the gate is made negative to the anode thus forward biasing the base–emitter junction of the p-n-p transistor TR1.

As the forward bias voltage applied to the gate is increased the gate current rises and the effect is to cause the SCR to turn 'on' at a lower forward break-over voltage as illustrated in Fig. 6.5. Once the SCR has 'fired' the gate voltage may be removed and the device will remain in the 'on' state. To switch the SCR 'off' it is necessary to remove or reduce the anode voltage so that the current falls below the 'holding level' I_h. The SCR will then revert to the forward blocking or 'off' state. The magnitude of the gate current required to switch-on the SCR is much smaller than the maximum anode-to-cathode current and since the gate–cathode voltage is small, the power that is supplied to the gate circuit need be only very small compared with the power that can be controlled by the SCR. Only a short pulse of current is required to switch an SCR 'on' and the time of switch-on is a few microseconds depending upon the magnitude of the gate current.

Although light may easily be excluded from the device, mechanism (c) permits an SCR to be switched 'on' by a beam of light. Such a device is called a Light Activated Silicon Controlled Rectifier (LASCR) which does not require a gate connection.

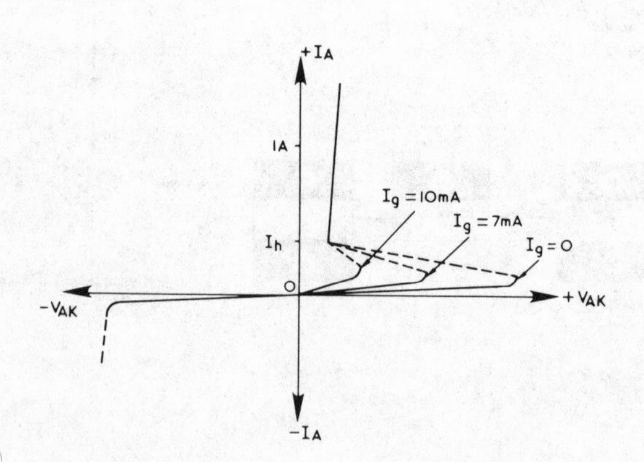

Fig. 6.5 Effect of increasing gate current.

Use of the SCR

The main use of an SCR is as a control device, controlling the amount of power fed to a load. Applications include the control of the average power fed to motors, lamps, heaters and d.c. supplies for electronic equipment.

The basic idea of one form of control is shown in Fig. 6.6 using a cathode-controlled SCR. The resistor *R* represents the load and the SCR is in series with it. The gate electrode is fed with positive pulses, the phase of which, relative to the a.c. supply, are made variable. If the phase of the pulses applied to the gate occur slightly delayed on the start of the positive half-cycles of the input voltage, the SCR will not fire until instants Z. Current will commence to flow in the load at these points and continue until instants X when the anode-to-cathode voltage has been reduced to zero. The magnitude of the current flowing during the 'on' period of the SCR will be determined by the value of *R* and the amplitude of the applied voltage. No current will flow during the negative half-cycles of the input as the SCR will be in the non-conducting direction. Current thus flows in bursts during each positive half-cycle. By delaying the phase of the gate triggering pulses, the SCR is made to fire at later instants Y in each positive half-cycle. As a result the current in the load flows in much shorter bursts as shown. The average value of the load current is proportional to the shaded area of the current wave form and can be varied from full value (corresponding to flow during a complete half-cycle) down to zero.

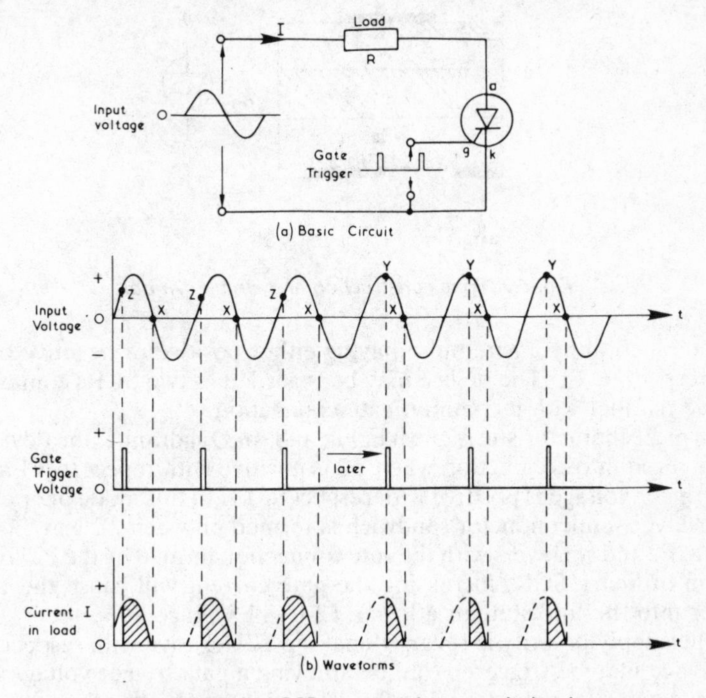

Fig. 6.6 Use of SCR to control current in load.

This method of controlling the average power fed to a load is known as 'phase control'. It is an efficient method when compared with traditional methods of controlling the power fed to lamps and motors using dropper resistors, gear boxes, clutches and drive belts, etc. If R is a lamp or motor the thermal or mechanical inertia of the load is used to smooth out the effect of the current pulsations.

The voltage drop across the SCR when 'on' is low (about 1V) hence the power loss is small compared with the power it is controlling. SCRs are available with forward and reverse blocking voltages up to 1kV and with current ratings up to several hundred amperes. Although the forward voltage drop is small, there is considerable power developed in the SCR with large-current operation and a heat sink must be used.

(b) Triac or Bidirectional SCR

A triac is a bidirectional triode thyristor and a basic construction is shown in Fig. 6.7. It has two main terminals T2 and T1 between which the main 'on' current flows and a gate electrode. The triac differs from the ordinary SCR in that current can be made to flow in either direction between T2 and T1, i.e. when T2 is positive to T1 or T2 negative to T1. In addition, it can be triggered

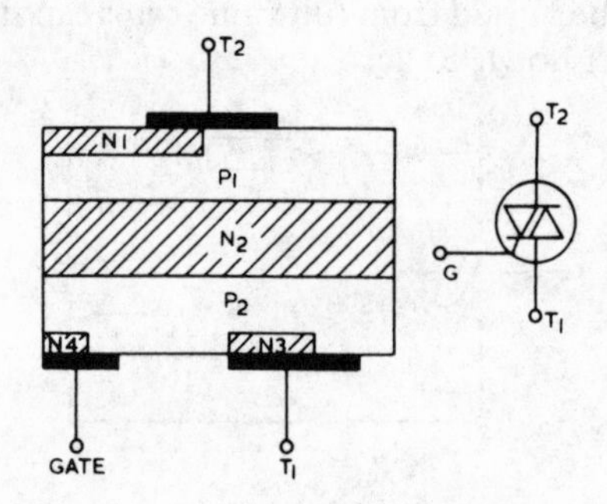

Fig. 6.7 Triac construction and circuit symbol.

into conduction by a gate pulse having either positive or negative polarity with respect to T1. The device may be regarded as two SCRs connected in inverse parallel with a common gate connection.

A typical characteristic is given in Fig. 6.8. In Quadrant 1, the device may be triggered into conduction when T2 is positive with repect to T1 and the gate trigger voltage is positive with respect to T1. In this mode of operation, a four-layer semiconductor sandwich is formed between T2 and T1 by the P1, N2, P2 and N3 layers with the gate connection formed by the P2 layer. As for the ordinary SCR, increasing the gate current will cause the triac to trigger into the 'on' state at a lower T2-to-T1 voltage.

If the supply polarity is reversed making T2 negative with respect to T1, the device may be triggered 'on' by applying a gate trigger voltage having negative polarity with respect to T1. This is shown by the characteristic in

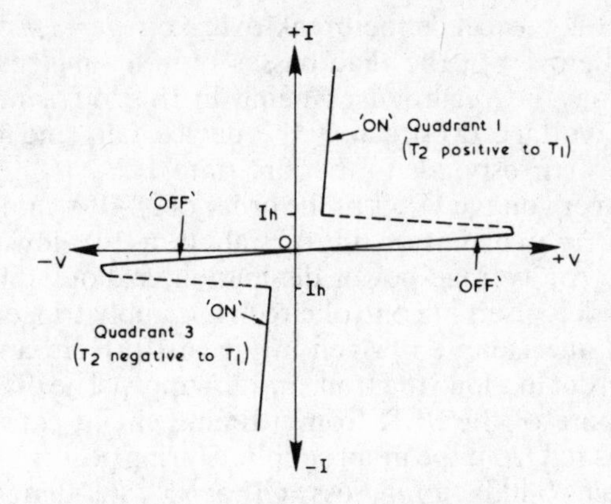

Fig. 6.8 Characteristic of the triac.

Quadrant 3. Here the four-layer sandwich is formed by the N1, P1, N2 and P2 layers between T2 and T1 with the gate connection formed by the N4 layer. Again, increasing the gate current will cause the triac to fire at a reduced value of voltage between T2 and T1.

To turn 'off' the triac in either of its 'on' states it is necessary to reduce or remove the voltage applied between the two main terminals so that the current falls below the holding level I_h. The advantage of the triac over the SCR is that it can be triggered into conduction with a supply voltage of either polarity applied to the main terminals and can thus be used in relatively simple circuits for the control of current in a.c. systems. In general, the current and voltage ratings of a triac are lower than those of an SCR.

(c) Diac

Diacs are bidirecitonal diode thyristors having two electrodes. A typical characteristic is shown in Fig. 6.9. The device can be made to conduct in either direction (with terminal A positive to terminal B or vice versa) when

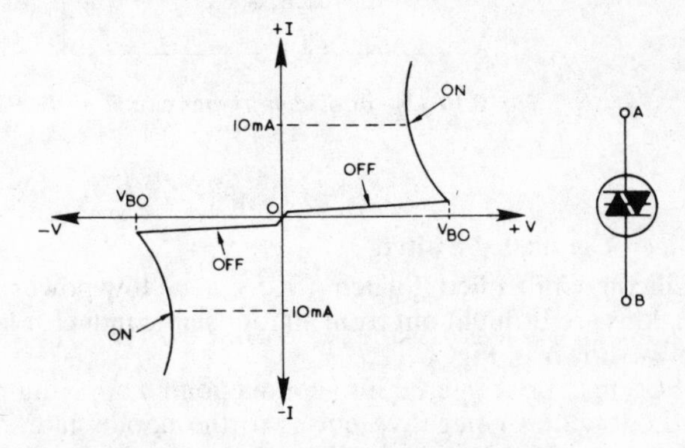

Fig. 6.9 Typical diac characteristic and circuit symbol.

the applied voltage exceeds the break-over voltage V_{bo}. With the applied voltage level below V_{bo} the diac passes only a small current (less than 100μA) and may be regarded as being in the 'off' state. When V_{bo} is exceeded, the voltage drop across the device falls and the current rises rapidly and this corresponds to the 'on' state.

The break-over voltage V_{bo} is of the order of 28–36V and the voltage drop across the diac at a current of 10mA is about 6–10V lower than V_{bo} for a typical device. An average power dissipation is about 150mW.

Diacs are widely used in control circuits to apply trigger pulses to SCRs and triacs and an example is given in Fig. 6.10. Here an SCR is used to control the current in a load fed from an a.c. supply. The diac supplies trigger pulses to the gate of the SCR from a timing circuit consisting of *R*1, *R*2 and *C*1 which is fed from the input supply. During positive half-cycles of the input a voltage builds up across *C*1 at a rate depending upon the time-constant of the timing circuit. When the voltage across *C*1 exceeds the break-over voltage of the diac it is triggered into conduction. *C*1 is then partially discharged by the gate current of the SCR which switches 'on' allowing current to flow in the load. The SCR switches 'off' in the usual way when its anode-to-cathode voltage falls to zero. During negative half-cycles of the input the voltage across *C*1 will reverse its polarity but the SCR will not come 'on' as it will be in its reverse blocking state. If a triac is used in place of the SCR, current will also flow in the load during negative half-cycles of the input when the reversal of voltage across *C*1 once exceeding V_{bo} will cause the diac to conduct in the other direction. Current will then flow in the gate of the triac causing it to go into the 'on' state (Quadrant 3 mode).

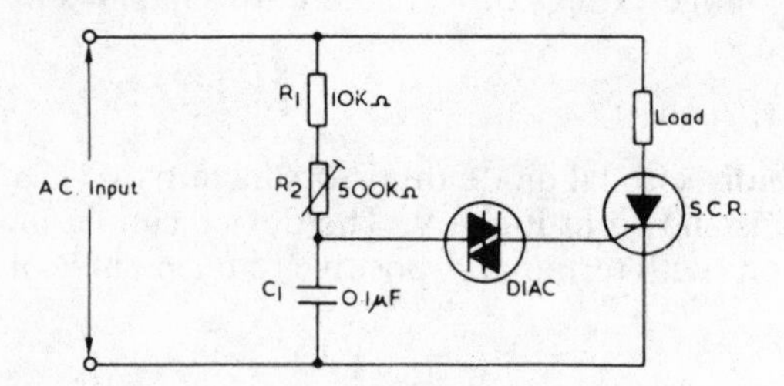

Fig. 6.10 Use of diac in trigger circuit for SCR.

(d) Silicon Controlled Switch

A Silicon Controlled Switch (SCS) is a low-power thyristor where connections are brought out from all four semiconductor layers. The circuit symbol is shown in Fig. 6.11.

An SCS may be triggered into conduction by applying positive pulses to the cathode gate or negative pulses to the anode gate. The operation is identical to an SCR but the additional gate improves the versatility of the

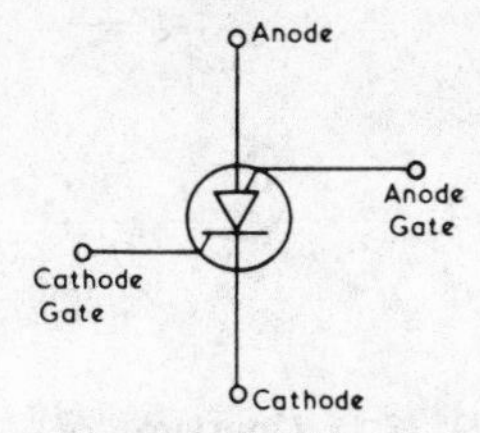

Fig. 6.11 Silicon controlled switch (SCS).

device. It is used in a wide range of switching applications including memory circuits and field oscillator circuits in television receivers.

UNIJUNCTION TRANSISTOR

The unijunction transistor (UJT) is another type of trigger device which is often used to control SCRs. The device consists of a small bar of n-type silicon with ohmic (non-rectifying) contacts at either end called the base connections B2 and B1, see Fig. 6.12. A p-n junction is formed between the emitter lead and the bar by alloying an aluminium wire to the bar as shown.

In use, B2 is supplied with a positive voltage with respect to B1 and in the absence of any voltage on the emitter, the resistance between B2 and B1 is high (of the order of 5–10kΩ). It may be considered that the silicon bar acts as a potential divider causing the N-region close to the p-n junction formed to take up a potential between that of B2 and B1. The actual potential depends upon the position of the emitter lead which is usually arranged so that the potential lies in the range of 0·4 to 0·8 of the voltage applied between B2 and B1.

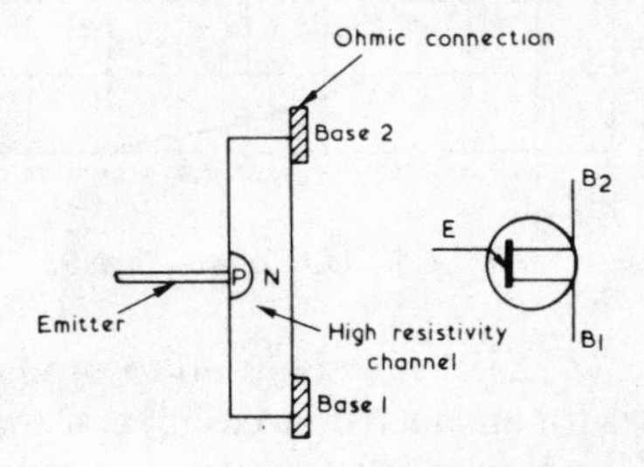

Fig. 6.12 Construction of unijunction transistor.

Suppose, for example, the voltage applied between B2 and B1 is 20V as shown in Fig. 6.13 and that the emitter is positioned such that the potential of the N-region close to the p-n junction is 14V (0·7 × the voltage between B2 and B1). Now, if a voltage V_{E-B1} is applied between emitter and B1 with polarity shown and is less than 14V, then the p-n junction will be reverse biased and no emitter current will flow. However, increasing V_{E-B1} so that it exceeds 14V by a small amount (say 0·6V) will cause the p-n junction to

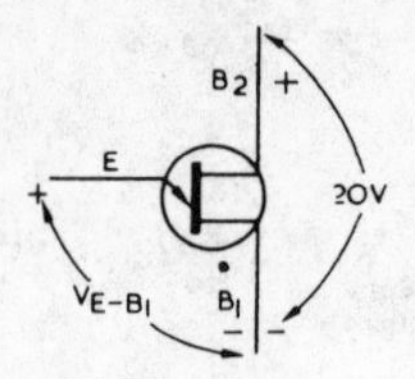

Fig. 6.13 Operation of UJT.

become forward biased and emitter current to flow. The flow of current between the emitter and B1 will reduce the resistance between these electrodes causing more current to flow. Thus the emitter current rises rapidly and the emitter-to-B1 voltage drops, see characteristic of Fig. 6.14. The characteristic shows that emitter current commences to flow when V_{E-B1} is approximately 14·6V but this will vary for a particular device depending upon the position of the emitter lead and the voltage applied between B2 and B1. After passing through a negative resistance region the current rapidly builds up and above a particular emitter current the voltage drop between emitter and B1 rises slightly.

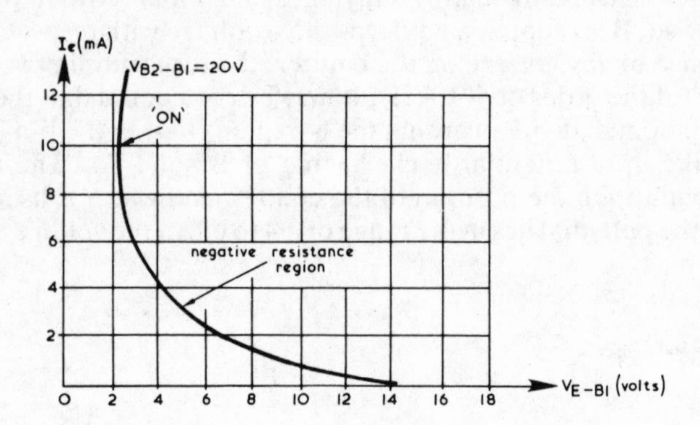

Fig. 6.14 UJT characteristic.

An example of how a UJT may be used to produce trigger pulses for an SCR or timing pulses for an electronic circuit is shown in Fig. 6.15. The UJT is arranged as a relaxation oscillator which generates a sawtooth voltage waveform at the emitter by the slow charging of a capacitor $C1$ to the supply voltage via $R1$ and $R2$ and the rapid discharging of $C1$ via the UJT.

At switch-on, $C1$ is uncharged, so the emitter potential is zero and the UJT is 'off'. As $C1$ slowly charges via $R1$ and $R2$, the voltage across $C1$ rises exponentially. When the voltage across $C1$ exceeds the potential of the N-region close to the p-n junction, the junction becomes forward biased causing emitter current to flow. $C1$ will therefore rapidly discharge via the low-value resistor $R4$. At the end of the discharge period the UJT junction becomes reverse biased and the device goes 'off'. $C1$ then recommences

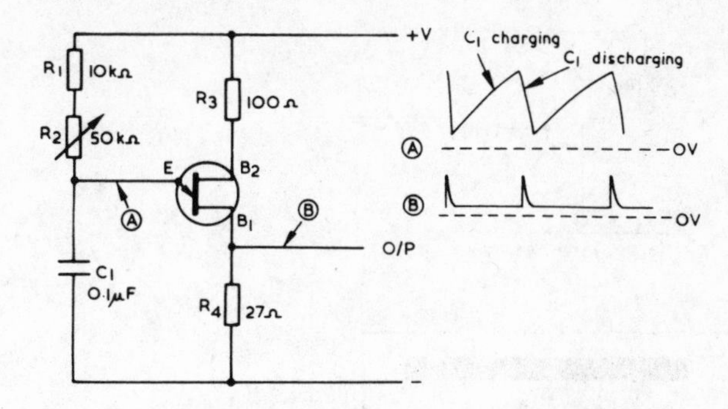

Fig. 6.15 UJT oscillator for producing timing pulses.

charging via *R*1 and *R*2 and the cycle is repeated. Each time *C*1 discharges, the current in *R*4 produces a positive-going pulse at the output. These pulses could be used to trigger an SCR or be used as timing pulses for some other purpose. Varying the *CR* time-constant by altering the value of *R*2 will vary the repetition rate of the output pulses. For the values given, the repetition rate will be in the range 30–200Hz depending upon the characteristics of the UJT. The resistor *R*3 is used to compensate for temperature variations.

MICROMINIATURE ELECTRONIC CIRCUITS

Following the introduction of the bipolar transistor into the electronics industry, efforts were made to make complete electronic circuits very small and more reliable. There are three types of microminature electronic circuits.

(1) Thin-film Circuits

With this type, the passive components, e.g. resistors, capacitors and conductors are built up by the evaporation of metals and insulators through masks to form patterns on a small piece of glass substrate.

The stages in the production of a thin-film circuit may be explained using the simple amplifier circuit of Fig. 6.16(a) and its thin-film version in Fig. 6.16(b). The first step is the evaporation of nichrome through a mask to form the resistors *R*1 and *R*2 on the glass substrate. The next stage is the evaporation of aluminium through a mask to form the lower electrode of the capacitor *C*1. This is followed by the evaporation of an insulator such as silicon monoxide to form the dielectric of *C*1, again applied through a suitable mask. The upper electrode of *C*1 is then produced by forming another thin film of aluminium through a mask on top of the silicon monoxide. Conducting strips are then formed to connect the thin-film components together by evaporating a metal such as gold through a suitable mask. Finally, the transistor is soldered into circuit, its leads being

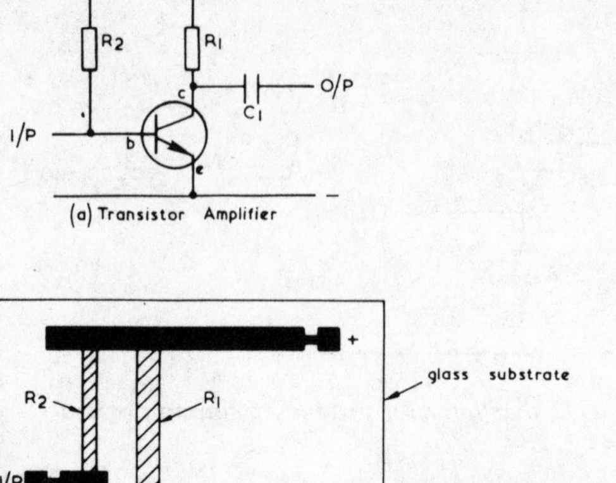

Fig. 6.16 Evaporated thin-film circuit.

connected to appropriate points on the conducting strips. The complete circuit is then encapsulated in epoxy resin cushioned by silicon rubber.

There are limits to the values of resistors and capacitors that can be formed by this technique and generally inductors cannot be produced. Evaporation of the materials on to the substrate (ceramic may be used in place of glass) is carried out in a high vacuum.

(2) Thick-film Circuits

These are similar to the thin-film circuits but are manufactured using 'screening' techniques. The passive components and conductors are formed by passing the resistive or conductive materials in the form of a glass mixture (frit) through a fine wire mesh to form patterns on a ceramic substrate. After drying, the ceramic substrate is heated in an oven to a high temperature (700–800°C). This causes the glass frit to melt and for the resistive or conductive material contained in the frit to form firm patterns on the substrate. On removal from the oven, hard glassy patterns are formed on the substrate which will withstand rougher handling than the fragile thin film circuits. The process of forming a complete microcircuit is carried out in stages using different frits for the resistors, conductors and capacitor component parts. Active devices (transistors and diodes) are soldered into circuit at points which have been previously tinned during the manufacturing stages of the substrate.

(3) Monolithic Silicon Integrated Circuits

These are the most important and most widely used type of microcircuit. Here, active and passive components are formed simultaneously on a small slice of a silicon crystal to create complete electronic circuits. Each complete microcircuit is approximately 1mm square and 0·3mm thick and up to one thousand can be made at a time from a thin slice of silicon having a diameter of 50mm, see Fig. 6.17.

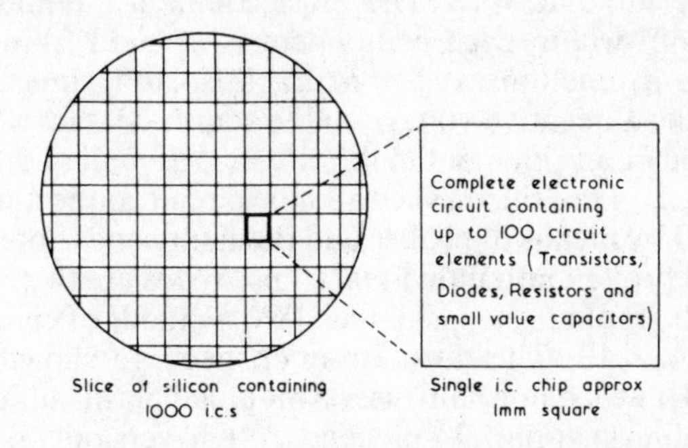

Fig. 6.17 An integrated circuit chip.

In one type of integrated circuit, bipolar transistor technology is used where transistors, diodes, resistors and small-value capacitors are formed by diffusing impurities into the silicon slice using techniques similar to those described for planar transistors in Chapter 3. All of the components are formed on the top surface of the silicon slice as shown in Fig. 6.18, and are interconnected to form the required circuit by evaporating aluminium connections on top of the silicon dioxide. Resistors can be made using a doped portion of the silicon as a resistance, the value depending upon the doping level, the length and cross-sectional area. These are called 'diffused resistors' and can be economically produced in the range 20Ω to 20kΩ. Resistor tolerances are about ±10% but it is possible to produce a better matching accuracy between two resistors on the same slice of ±2%. Thus accurate potential dividers can be formed using resistance ratio rather than absolute values as the controlling factor.

One way of forming an integrated capacitor is to utilise the capacitance of a reverse biased p-n junction but this will require a suitable reverse bias

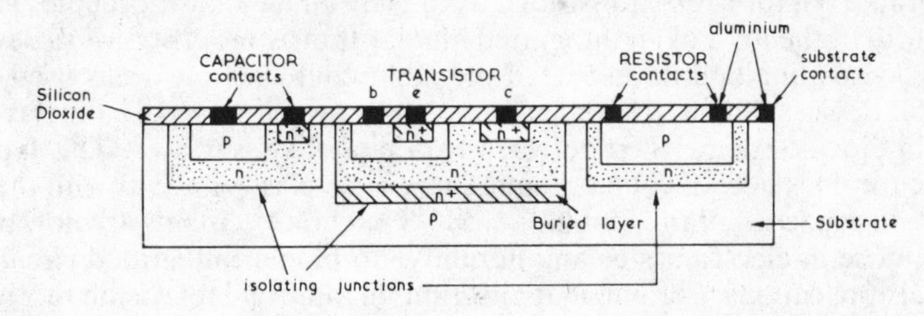

Fig. 6.18 Components in integrated circuit form (bipolar transistor technology).

voltage. Capacitance values up to about 100pF can be obtained by this technique. If larger values are required, discrete capacitors must be connected external to the i.c. In Fig. 6.18 the resistor element is formed by the p-channel between the contacts and the capacitor utilises the capacitance of a p-n junction. The transistor is an n-p-n using a special buried layer (n^+) to reduce the collector resistance. As all of the components on an i.c. are formed on the same slice of a conductive substrate, they must be isolated from each other in some way. The most common method is known as 'junction isolation' where each component is isolated from the common substrate by a p-n junction as shown. Each isolating junction is reverse biased by applying a negative voltage to the source contact which is greater than the potential of any element in the circuit. After all of the components have been formed on the silicon slice and the circuit connections deposited, each i.c. is tested by means of probes and the faulty ones noted. The slice is then scribed and broken into individual chips. Wires are then bonded to the good chips which are then encapsulated. Two methods of encapsulation are illustrated in Fig. 6.19. The most common package shown in (a) is the **dual-in-line** (DIL) which normally uses a moulded plastic unit. Dual-in-line types are available in 8-pin, 14-pin and 16-pin versions. The TO metal package of (b) is similar to that used for a discrete transistor and is available with 8 pins, 10 pins or 12 pins.

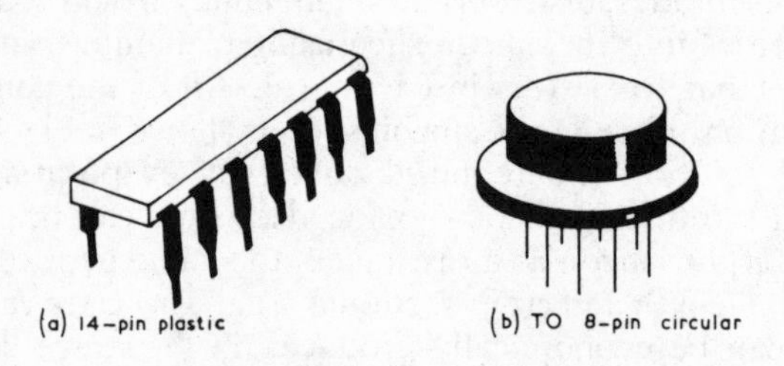

Fig. 6.19 Integrated circuit packages.

The metal-oxide semiconductor technology used in the manufacture of field effect transistors can also be employed to produce complete circuits in integrated circuit form. These i.c.s are called 'MOS integrated circuits'. An integrated circuit MOS transistor has the advantage that it occupies about one-fifth of the area of an integrated bipolar transistor. Also, MOS devices are self-isolating and do not need the additional isolating areas used with bipolar devices. Small MOS capacitors can be produced using a thin layer of silicon dioxide sandwiched between two 'plate-like' surfaces. This type of capacitor does not, of course, require a reverse bias voltage as with the p-n junction capacitor of the bipolar i.c. MOS integrated circuits are now being used more in electronics as an alternative to bipolar integrated circuits in digital applications. For linear applications in radio and television receivers, bipolar transistor i.c. technology is the most common.

HANDLING INTEGRATED CIRCUITS

The following precautions should be observed when handling integrated circuits.

(a) Soldering-iron temperature. An i.c. may be damaged permanently if excess heat is transferred to it during soldering or desoldering operations. With the tip of the soldering iron at a temperature in the range 245°C to 400°C the soldering or desoldering time should be less than 5 secs. At a tip temperature below 245°C the maximum time may be increased to 10 secs. Soldering irons with thermostatically controlled element temperatures are available to meet the special temperature requirements of i.c.s.

(b) Removal of I.C. Where an i.c. is soldered on to a printed circuit board it may be removed by desoldering each pin in turn. To reduce the risk of excess heat being transferred to the i.c. the molten solder may be quickly removed by suction using a desoldering tool. Integrated circuits which plug into a socket must be removed carefully to avoid damage to the pins which are brittle. Uneven grip during extraction may cause the pins to bend, twist or fracture. Removal is best carried out using an extraction tool which has specially designed teeth to grip the i.c. during and after extraction.

(c) Fitting of I.C. Before fitting a new i.c. into circuit it is prudent to check that the d.c. supplies and input signal levels are of normal magnitude, also that the external components are not faulty e.g. short circuit, low or high resistance. Note that excess voltage or faulty components may damage the i.c. Next, check that the i.c. is the right way round; a slot or white spot at one end of the package is normally used as a reference to ensure correct orientation. Prior to insertion the pins of the i.c. should be carefully aligned. An i.c. insertion tool is particularly useful in this respect. Not only does the tool ensure that even pressure is applied when inserting the i.c. but automatically aligns the pins to the correct insertion distance.

(d) Checking of I.C. When checking d.c. voltages or waveforms on the pins of an i.c. ensure that the test probe does not short pins together otherwise permanent damage may occur. It may be rather difficult in some items of electronic equipment to get to the pins on both sides of the i.c. In these circumstances an i.c. test clip should be used. This is clipped over the i.c. to be tested and brings out the pin connections on top where there is less risk of accidental shorting.

(e) Static Charges. MOS integrated circuits, like MOS field effect transistors, can be damaged by electrostatic discharges. Although MOS i.c.s incorporate built-in protection against the effect of static charges, they can nevertheless be damaged by over-voltages. For this reason MOS devices are usually supplied in conductive bags which should not be removed during storing or transporting.

Special precautions should be taken when handling as static may be introduced during normal servicing. Manufacturers recommend that work on MOS devices be carried out on a conductive surface, e.g. a metal top bench and that the person doing the servicing be connected to the conductive surface by a metal bracelet and a conductive cord or chain. Everything connected with the servicing operation should be at the same potential as the conductive surface, e.g. tools, printed board receiving the i.c., the i.c. and the person carrying out the work. It is important, however, that in conforming to recommended procedures, personal safety is not endangered.

MOS devices should not be inserted or removed whilst the equipment is switched 'on', nor signals applied to the i.c. when the power is 'off'.

LIGHT-EMITTING DIODES

All p-n diodes passing current in the forward direction emit radiation. This radiation results from the recombination of minority carriers with majority carriers in the vicinity of the junction and, in simple terms, is due to electrons moving from a high-energy level to a lower-energy level. In moving from a high-energy level to a lower one, electrons give up energy in the form of an e.m. radiation.

The wavelength of the emitted radiation depends upon the size of the 'band gap' of the material used in the construction of the diode, see Fig. 6.20. To obtain emission in the visible light spectrum, band gaps in excess of 1·71eV are required. This rules out the common semiconductor materials of silicon and germanium whose band gaps are 1·1eV and 0·7eV respectively. A suitable crystalline material is gallium arsenide phosphide which can be made to emit light in the red, green, yellow or infra-red parts of the e.m. spectrum.

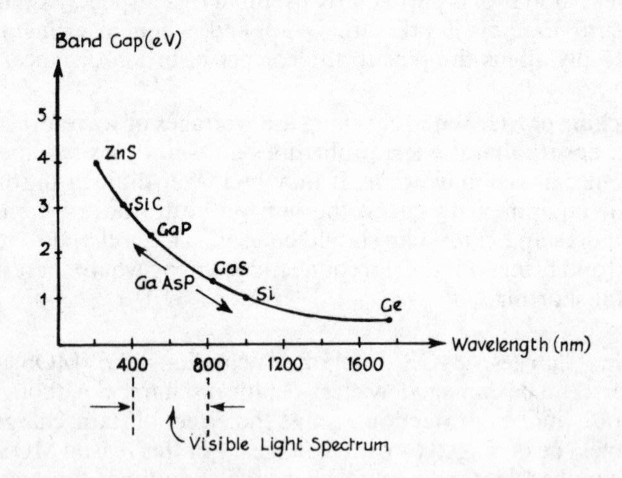

Fig. 6.20 Wavelength of radiation emitted from forward biased p-n junction.

Not all of the light that is radiated from the vicinity of the junction finds its way out of the surface of the material, see Fig. 6.21(a). Some of the photons are absorbed and some are totally internally reflected. The internal reflection can be reduced by placing the crystal in intimate contact with an epoxy plastic lens of high refractive index, see Fig. 6.21(b).

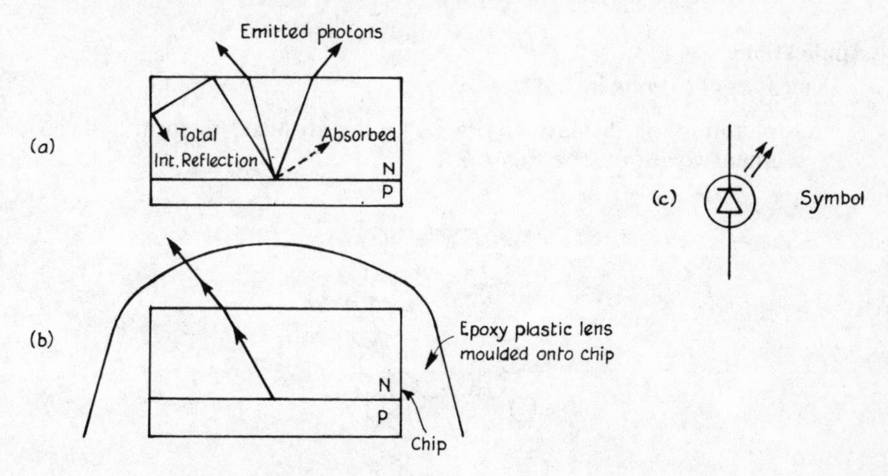

Fig. 6.21 Use of integral lens to reduce loss due to total internal reflection.

Voltage–Current Characteristic

Electrically, light-emitting diodes resemble ordinary junction diodes. Those made from gallium arsenide and gallium phosphide have higher forward voltage drops than silicon diodes (1·5V to 2·0V), see Fig. 6.22. They have lower reverse voltages (3V to 10V) but similar temperature coefficients (−2mV/°C at room temperature).

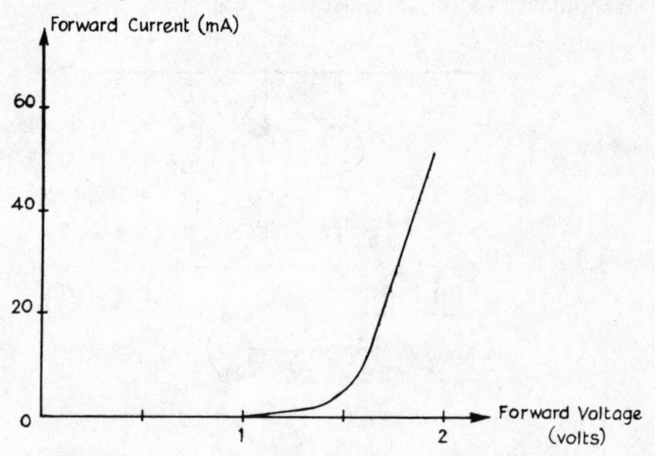

Fig. 6.22 Forward characteristics of l.e.d.

Efficiency of radiation increases with forward current; however, this is more than offset by the consequent rise in junction temperature, when the efficiency falls. By operating an l.e.d. under pulse conditions, the high efficiency can be used to advantage since the rise in temperature is minimised because of the cooling period between pulses.

Applications

Typical applications include:

(a) Alpha-numeric displays where 28 l.e.d.s are used to form a seven-segment read-out, see Fig. 6.23.

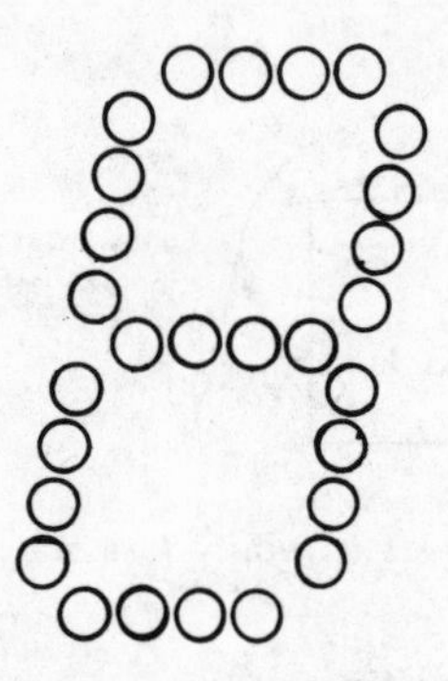

Fig. 6.23 28 L.E.D.s forming seven-segment read-out.

(b) Tuning indicators for f.m. receivers, giving an indication of signal strength and stereo signal operation.

(c) Infra-red remote control transmitters where an infra-red beam is modulated by digital pulses, see Fig. 6.24.

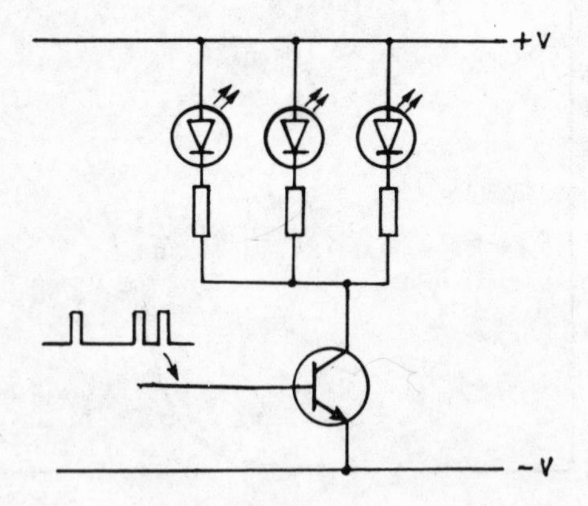

Fig. 6.24 Part of infra-red transmitter showing pulse operation of l.e.d.s.

LIQUID CRYSTAL DISPLAY

An alternative to an l.e.d. display is one which uses a **liquid crystal**. The basic idea of the common form of liquid-crystal display is given in Fig. 6.25(a) which illustrates the construction of a seven-segment element. Here liquid-crystal material is sandwiched between two sheets of glass which have conductive coatings on their insides. One glass plate has a common electrode etched on it. These very thin conductive coatings (tin oxide) will allow the transmission of light through them.

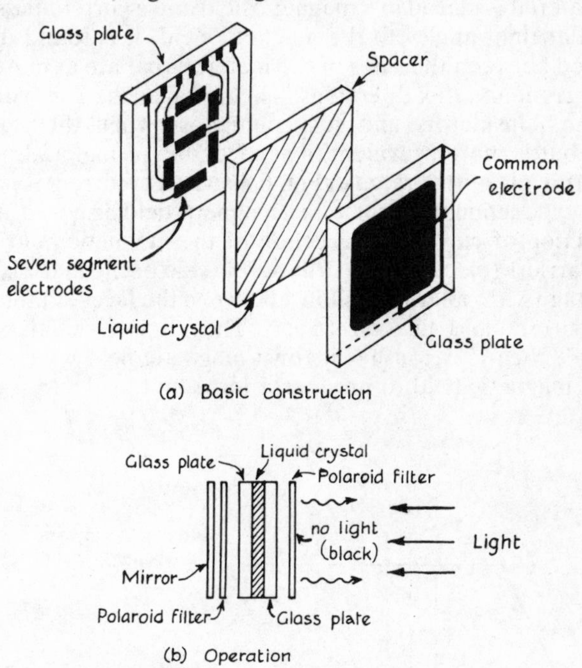

Fig. 6.25 Liquid crystal seven-segment display.

To produce a particular display digit, a voltage is applied between the appropriate segments and the common electrode. This causes a change in the optical properties of the liquid crystal which appears as the display. The operation is illustrated in Fig. 6.25(b), using additionally two polaroid filters (with their planes of polarisation at right-angles to one another) and a mirror. Without any voltage applied between the electrodes, external light falling on the device is polarised by the first filter and is twisted through 90° by the liquid crystal so that it will pass through the second filter. After reflection by the mirror the light returns by the same path; in this state the cell is 'clear'.

When a voltage is applied between any of the segments and the common electrode the optical properties of the liquid crystal change in those areas. The liquid crystal then no longer produces the 90° twisting of the polarised light as it enters the crystal, thus the light will not pass the second polaroid filter. Hence the parts of the display where the electric field is applied appear black.

L.E.D. and liquid-crystal display drivers and decoders are considered in Volume 3 of this series.

HALL-EFFECT DEVICES

If a material is placed in a magnetic field and a current passed through the material at right-angles to the magnetic field, it is found that a voltage is developed between the sides of the material that are at mutual right-angles to the current and flux directions. This voltage (the Hall voltage) occurs in conductors, insulators and semiconductors when the stated conditions prevail, but is more prevalent in semiconductor materials.

The principle is illustrated in Fig. 6.26 where a current is passed through a slab of n-type semiconductor with a magnetic field disposed at right-angles to the direction of current. The action of the magnetic field is to cause the charge carriers (electrons) to drift to one side of the material, resulting in a Hall voltage with polarity as shown between the faces at mutual right-angles to the current and flux directions. The effect is similar to the 'motor principle'; each electron has its own magnetic field which reacts with the external magnetic field to produce a force on the electron causing it to be

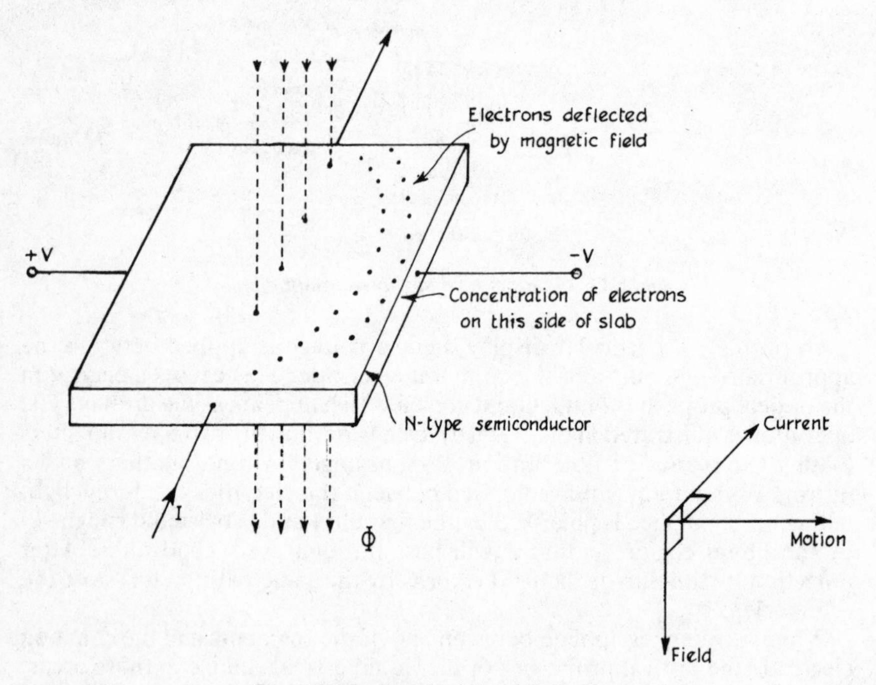

Fig. 6.26 Basic principle of Hall effect.

deflected. If the material were p-type, the charge carriers (holes) would drift to the right and the polarity of the Hall voltage would reverse.

The magnitude of the Hall voltage generated is small (tens of millivolts) and is proportional to the flux density of the magnetic field and the strength of the current that is passed through the material. Also, reversing the direction of the current or the magnetic field will reverse the polarity of the Hall voltage.

Applications

The Hall-effect device (small slab of semiconductor material) is usually mounted in an i.c. together with amplifiers to amplify the small Hall voltage that is generated. Connections are brought out so that an external voltage may be supplied to produce the current to be passed through the semiconductor material, the same voltage normally being used to provide the supply for the i.c. amplifier. The magnetic field is supplied externally. Typical applications include:

(a) Magnetic Field Measurements

The Hall effect device is mounted at the end of a probe which is placed in the centre of the magnetic field to be measured. The Hall voltage that is generated is amplified and used to provide an indication on a meter which may be calibrated in units of magnetic flux density (tesla).

(b) Audio Power Measurements

In this application, the d.c. voltage source is replaced by an audio voltage source. The alternating Hall voltage produced is amplified and used to provide an indication on a meter which may be calibrated in watts.

(c) Proximity Switch

A Hall-effect device may be utilised to give a 'bounce-free' switching action when influenced by a magnetic field and the idea is shown in Fig. 6.27.

When the small permanent magnet is bought close to the magnetic centre of the Hall device i.c. a Hall voltage is generated. On reducing the proximity of the magnet and the i.c. by only a few millimetres the Hall voltage will

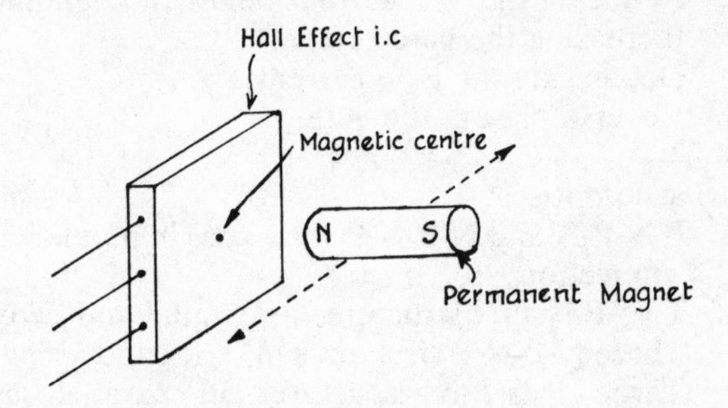

Fig. 6.27 Hall-effect proximity switch.

cease to be generated. Thus the output voltage from the i.c. may be used to provide a switching action and is particularly useful in logic circuits where 'bounce-free' switching is required.

A similar idea is used in the 'vane' switch illustrated in Fig. 6.28, but here the permanent magnet is in a fixed position so that a Hall voltage is generated continually by the i.c. sensor. However, when a ferrous metal vane passes through the gap between the magnet and the sensor the Hall voltage is inhibited thus providing a 'bounce-free' switch action. The device may be used for position or counting applications in industry, particularly in dusty or high ambient lighting situations where optical sensors would be unsuitable.

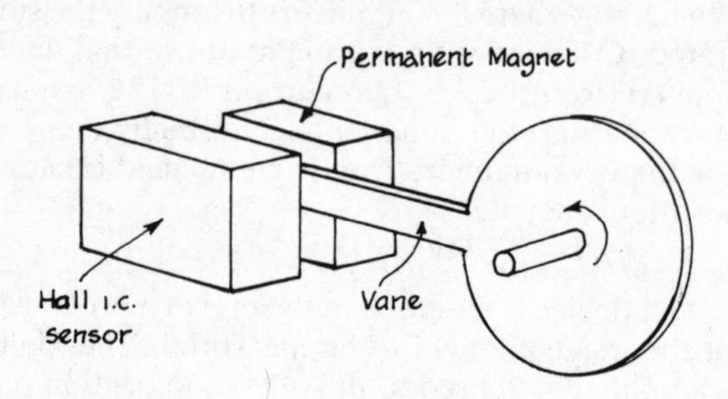

Fig. 6.28 Hall-effect vane switch.

QUESTIONS ON CHAPTER SIX

(1) An SCR has:
- (a) An anode and cathode only
- (b) An anode, cathode and two gates
- (c) A source, drain and one gate
- (d) An anode, cathode and a single gate.

(2) An SCR may be taken from the 'on' to the 'off' condition by:
- (a) Reducing the 'on' current below the 'holding' level
- (b) Increasing the gate current
- (c) Decreasing the gate current
- (d) Reverse biasing the gate.

(3) A triac comprises:
- (a) P-N-P-N and N-P-N-P structures in parallel and a common gate connection
- (b) Two P-N-P-N structures in parallel and two gate connections
- (c) Three N-P-N-P structures in series and a single gate connection
- (d) Three P-N-P-N structures in parallel and a single gate connection.

(4) The circuit symbol for a triac is:

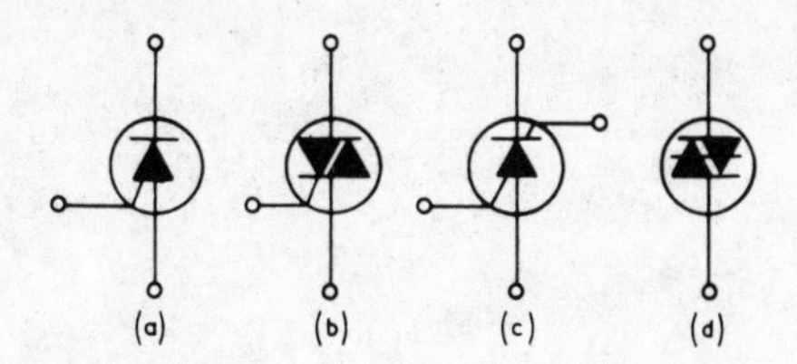

(5) A bidirectional thyristor having two electrodes only is called:
(a) A Unijunction Transistor
(b) A Diac
(c) A Silicon Controlled Switch
(d) An SCR.

(6) Microcircuits where complete electronic circuits are formed on a small chip of silicon are called:
(a) Printed circuits
(b) Monolithic integrated circuits
(c) Thick-film circuits
(d) Thin-film circuits.

(7) There is 24V applied between the B2 and B1 electrodes of a unijunction transistor and the emitter is positioned so that it lies three-quarters of the distance from B1 to B2. The device will be in the 'on' condition when the emitter-to-B1 voltage is approximately:
(a) 12·4V
(b) 18·6V
(c) 6·6V
(d) 16·6V.

(Answers on page 180)

CHAPTER SEVEN

THE CATHODE RAY TUBE

THE CATHODE RAY tube (c.r.t.) forms the basis of the display devices found in monochrome and colour television receivers, oscilloscopes and visual display units used in computers and other systems. In a c.r.t. a fine beam of electrons is directed at high velocity towards a glass screen the inside of which is coated with a layer of electroluminescent material which emits light on being struck by the electrons, see Fig. 7.1. This layer is called the **screen phosphor** and the colour of the light emitted depends upon the chemical composition of the layer. With cathode ray oscilloscopes the c.r.t. phosphor usually emits green or blue light. For a monochrome television tube a mixture of phosphors emitting blue and yellow light is used to create white light, whereas in a colour television tube separate phosphors in the form of small dots or stripes are used to produce light in the primary colours of red, green and blue. The emission of light is due to both fluorescence and phosphorescence. Fluorescence occurs when the screen layer is excited by the electrons whilst phosphorescence occurs after the excitation has ceased and is called the 'afterglow'.

The fine electron beam is produced by an electron gun assembly which is positioned inside a funnel-shape glass envelope that is highly evacuated, i.e. all the air and other gases released from the electrodes during manufacture are pumped out. It is normally arranged for the electron beam to be deflected either horizontally or vertically or in both directions simultaneously. The mechanism producing deflection has not been shown in Fig. 7.1 as the 'deflecting field' may originate from inside the neck of the

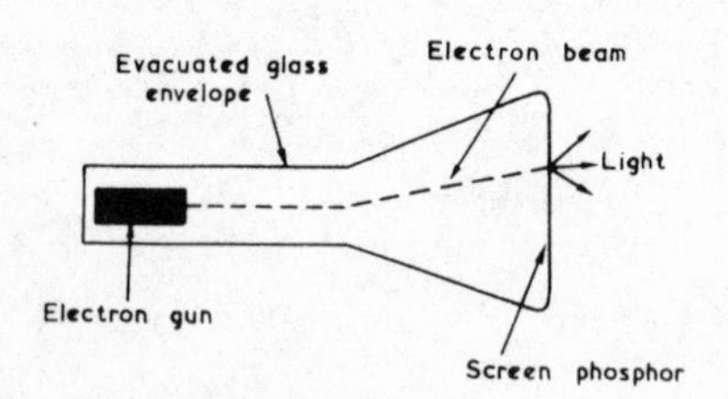

Fig. 7.1 Basic idea of c.r.t.

c.r.t. or outside it depending upon whether electric or magnetic deflection is used. A cathode ray tube can be divided into three basic sections.

(1) The electron gun assembly which produces the electrons and focuses them into a fine beam.
(2) The deflecting system which deflects the beam over the screen.
(3) The viewing area or screen which emits light on being excited by the arriving electrons.

The basic principle and construction of an oscilloscope tube will be considered under these three headings.

(1) ELECTRON GUN

Before an electron beam can be produced a source of electrons is required. At normal room temperatures the electrons in a metal wander at random through the atomic structure of the material. The electrons have insufficient energy to leave the material owing to the electrostatic forces that restrain them. If, however, sufficient additional energy is supplied to the material, high-energy electrons near the surface overcome the retarding electrostatic forces and escape from the material. One way of providing the additional energy to cause the emission of electrons is by the application of heat. This process is called thermionic emission, see Fig. 7.2, and is the principle used in the cathode ray tube and other thermionic devices such as valves.

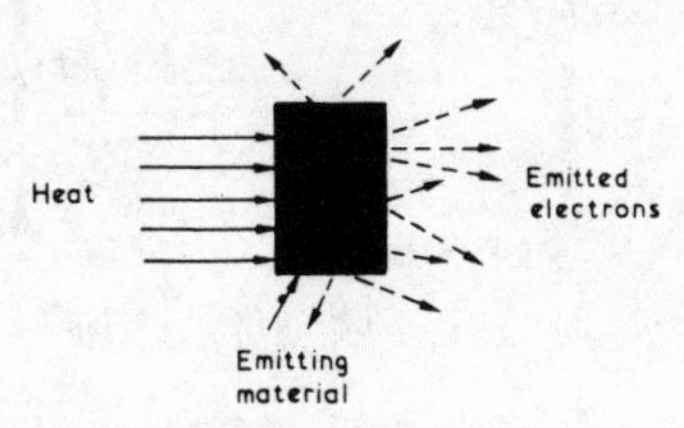

Fig. 7.2 Thermionic emission.

In practice three main materials are used as thermionic emitters but the most common and that used in c.r.t.s consits of a mixture of barium and strontium oxides – referred to simply as 'oxide emitters'. These oxides give appreciable emission at a temperature in the range 750°C – 950°C which corresponds to a red heat. Since the oxides are white powders it is necessary to apply them as a coating on a conductor, which can be suitably heated. The oxide coated conductor (usually nickel) is called the **cathode** of the electron gun assembly.

Two forms of cathode construction for use in a c.r.t. are shown in Fig. 7.3. In diagram (a) the cathode is a nickel tube with an oxide coating at one end. The oxide area is made as small as possible to produce a narrow electron beam source. To heat the oxide to its emitting temperature a tungsten heater wire is placed inside the nickel tube but is electrically insulated from it by

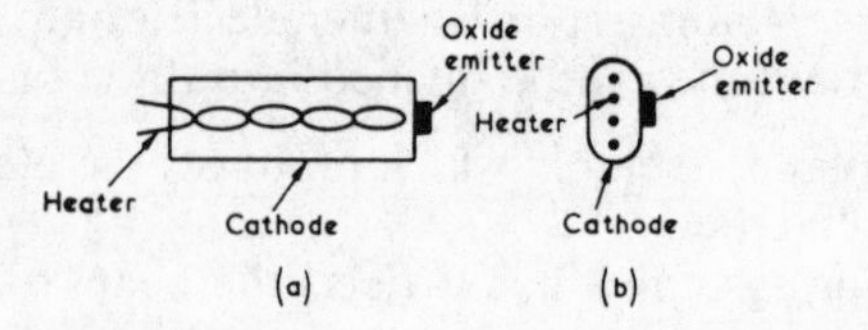

Fig. 7.3 C.R.T. cathode construction.

coating the heater with aluminium oxide. In diagram (b) the cathode is formed from a short length of flat tube containing the heater wire with again a small area of oxide coating on one side. When the cathode is raised to its emitting temperature by passing a current through the heater wire, electrons are emitted in all directions from the oxide but the velocity of the emitted electrons will not be high. The electrons must be concentrated into a fine beam and accelerated to a high velocity so that a bright enough spot is formed on the screen of the c.r.t.

A simple way of accelerating the emitted electrons to a high velocity and producing a beam is shown in Fig. 7.4(a). Here a plate called the **anode** with a small hole in it is placed between the cathode and the c.r.t. screen. If the

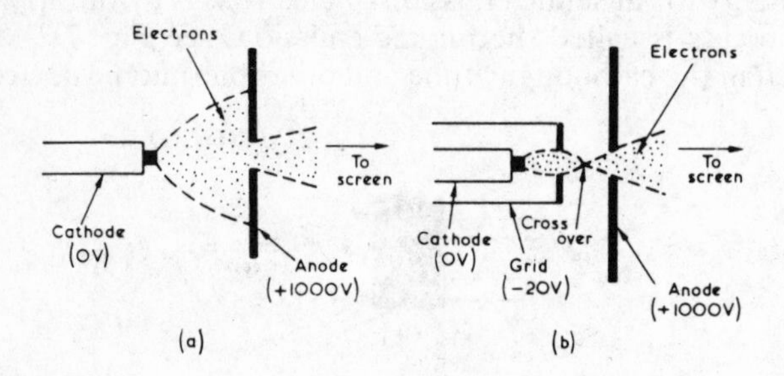

Fig. 7.4 Producing a beam of electrons.

anode plate is held positive with respect to the cathode the emitted electrons will be attracted towards it (owing to the attraction of unlike charges). The anode will collect a large number of electrons but some will pass through the hole to form a beam. The arrangement is not very efficient as it stands since only a small proportion of the emitted electrons passes through the hole. A better and more practical arrangement is shown in Fig. 7.4(b). A cylinder which is closed at one end except for a small hole at its centre is placed over the cathode. This cylinder is sometimes known as a 'Wehnelt cylinder' but is often referred to as the **grid**. The cylinder is placed at a negative potential with respect to the cathode so that it tends to repel the negative electrons (like charges repel). Electrons which are emitted from the cathode at an angle are now turned round and cross over as shown before passing through the anode aperture. We now have a more efficient arrangement having provided a more intense beam passing through the anode. The velocity of the electrons leaving the anode will depend upon the anode voltage and the higher the voltage the greater the velocity. With an anode voltage of 1000V the velocity will be approximately 19×10^6 metres per second which is very high. By varying the potential of the grid, the number of electrons in the beam or the beam current may be altered, hence adjusting the brightness of the spot on the screen. As the grid is made more negative to the cathode there is a greater repulsive force acting on the emitted electrons urging them back towards the cathode. The beam current is made smaller and the spot brightness reduced. If the grid is made sufficiently negative no electrons will pass through the grid aperture, hence the beam current will be zero and the spot of light extinguished.

Focusing the Beam

The divergent beam of electrons emanating from the anode aperture of Fig. 7.4(b) has to travel the length of the tube to the screen. During its travel there will be some additional widening of the beam due to the natural electrical repulsion between electrons in the beam. Thus the spot on the screen would be large and not very useful. It is therefore necessary to focus the beam so that a fine spot of light is produced on the screen. An **electrostatic lens** is used for this purpose which behaves in a way similar to an optical lens on a beam of light. One type of electrostatic lens is shown in Fig. 7.5 consisting of three anodes. The first anode, grid and cathode operate as previously described. The first and third anodes are in the form of discs whilst the second anode is usually of cylindrical construction. A potential of about one-fifth of the first and third anode voltage is applied to the second anode. In this example the five electrodes constitute the complete 'electron gun' assembly. Since the anodes are at different potentials, electric or electrostatic fields will be set between them as shown. Electrons entering these fields are subjected to forces urging them to travel in paths exactly opposite to the direction of the lines of force. When an electron enters a field at an angle, its direction will therefore be changed. This principle is the basis of the electrostatic lens. Due to the shape of the anodes and their electric field patterns a converging beam of electrons emerges from the third anode

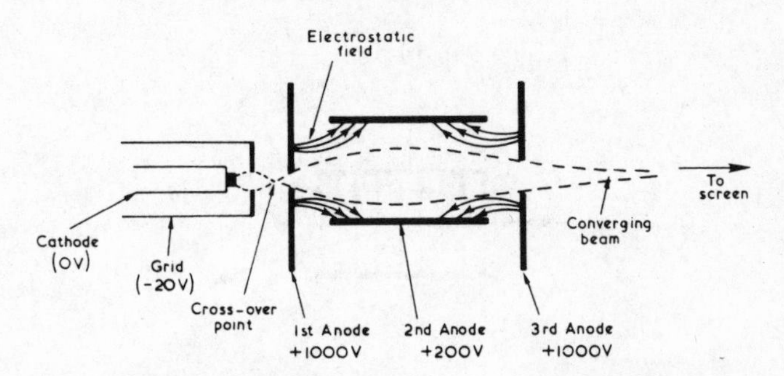

Fig. 7.5 Electrostatic focusing lens formed by three anodes.

to produce a fine spot of light on the c.r.t. screen. While within the electric fields the electons may travel in curved paths but on leaving the fields they travel in straight line paths. In effect the electrostatic lens focuses the crossover point which is the apparent source of electrons or 'object' on to the screen of the c.r.t.

To ensure that the point of focus coincides exactly with the screen of the c.r.t., the voltage fed to the second anode (called the 'focus anode') is made variable. This is equivalent to varying the focal length of an optical lens. Some electrostatic lenses use only two anodes but the principle of operation is the same.

(2) DEFLECTION SYSTEM

Now that the electron beam has been focused to produce a small spot of light on the c.r.t. screen, it is necessary to be able to deflect the beam in two directions at right angles. For cathode ray tubes used in oscilloscopes this is achieved by employing **electrostatic deflection**, the principle of which is shown in Fig. 7.6

The electron beam is passed through a pair of plates as shown in Fig. 7.6(a). If there is no voltage between the plates, the electrons will continue to travel in a straight line along the tube axis and the beam will not be deflected. With a voltage applied between the plates making the upper plate positive to the lower plate as shown, the electrons will be attracted towards the upper plate and repelled by the lower plate. Thus the electron beam will be bent upwards as it passes between the plates. The force acting on the beam while it is between the plates urges the electrons to travel in the exact opposite direction to the lines of the electrostatic field. This force progressively alters the path of the electrons from the point of entry to the point of exit causing them to follow a curved path in the deflecting field. On

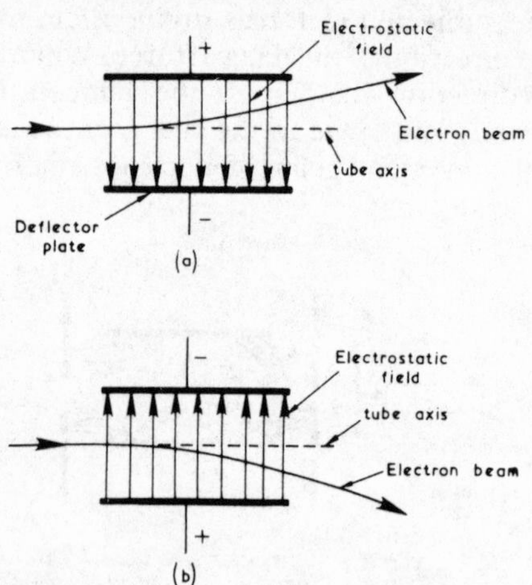

Fig. 7.6 Electrostatic deflection of electron beam.

leaving the deflecting field, the electrons travel in a straight line path at an angle to the tube axis. The spot on the screen will therefore be deflected in an upward direction. The amount of deflection, and hence the movement of the spot, is proportional to the magnitude of the voltage applied between the plates. The **deflection sensitivity** of the plates is the voltage that must be applied between the plates to produce 1cm deflection of the spot on the c.r.t. screen. If the polarity of the voltage applied between the plates is reversed as in Fig. 7.6(b) the beam and hence the spot on the screen will be deflected in a downward direction.

Since we require deflection of the beam in two directions at right-angles, two sets of deflector plates disposed at right-angles to one another are necessary as shown in Fig. 7.7. In an oscilloscope tube the beam is first

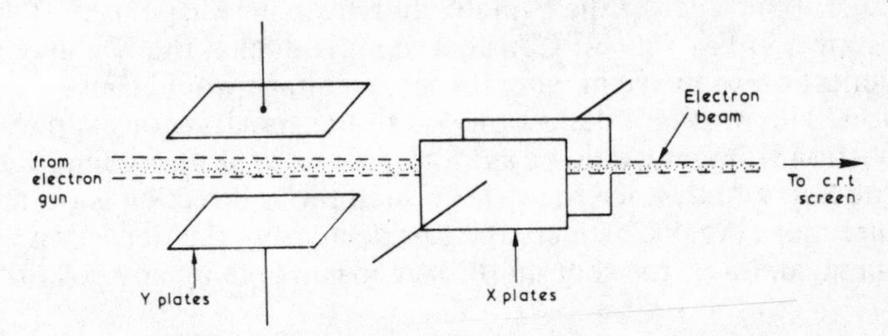

Fig. 7.7 Deflection of electron beam in two directions at right-angles.

deflected in the vertical or Y direction by the **Y-plates** and then deflected in the horizontal or X direction by the **X-plates**. The X-plates are supplied with a sawtooth voltage from the internal timebase oscillator of the c.r.o. while the Y-plates are fed with the signal to be examined. As quite large voltages (20 to 100V) are required to produce appreciable deflection of the beam, the plates are fed via suitable amplifiers. The plates nearest to the gun assembly produce the greatest deflection sensitivity (all other factors being equal) since the distance to the screen from these plates is longer than for the other set of plates. It is thus advantageous to feed the signal to be examined to the plates having the greatest deflection sensitivity. Usually the deflecting plates are shaped using one of the configurations shown in Fig. 7.8. This enables, for a given length of c.r.t., greater deflection angles to be obtained without significantly reducing the deflection sensitivity.

The use of two sets of plates at right-angles allows the beam to be deflected at any angle to the X or Y axis of the tube thereby permitting the spot of light

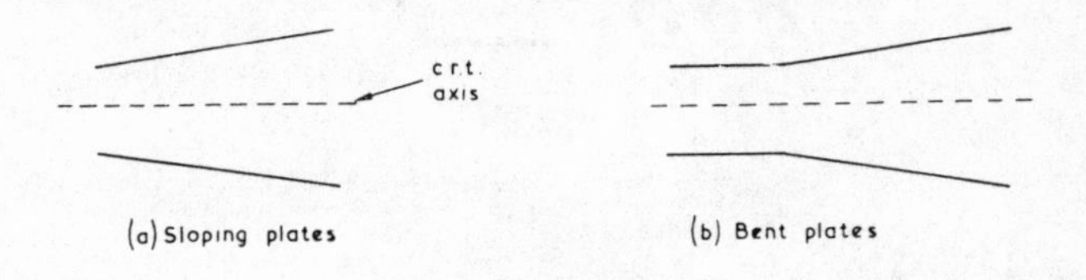

Fig. 7.8 Use of shaped deflecting plates.

to take up a position anywhere on the screen. Two examples are shown in Fig. 7.9 to illustrate this effect. In Fig. 7.9(a) equal steady voltages are applied to the two sets of plates (assumed to be of the same sensitivity) with polarities as indicated. Considered individually, the effect of the voltage applied between the X-plates would cause the beam to move to the right whereas the voltage applied between the Y-plates would cause the beam to more upwards. The resultant deflection of the beam with voltages applied simultaneously is as indicated and at 45° since the deflection voltages are the same. The spot on the screen can therefore take up any position along this resultant path by increasing or decreasing the deflection voltages by the same amount. If the polarities of both voltages are reversed, the resultant deflection will be diagonally opposite to that shown. The effect of reducing the voltage applied across the Y-plates and reversing the polarity at the same time is shown in Fig. 7.9(b). Considered individually, the X voltage would cause deflection to the right whereas the Y voltage would cause downward deflection. The resultant deflection due to the simultaneous application of the deflection voltages would be as indicated and at an approximate angle of 27° to the horizontal, since the deflection in the X direction is twice that in the Y direction. Again, by increasing or decreasing the deflection voltages by the same amount, the spot on the screen can take up any position along

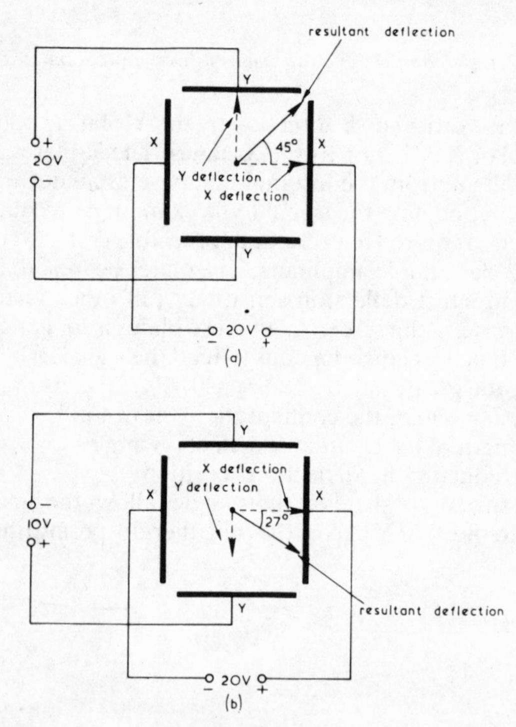

Fig. 7.9 Examples showing simultaneous deflection of beam in two directions at right-angles with steady voltages applied to the plates.

this resultant. If the polarities of both voltages are reversed the resultant deflection of the beam will be diagonally opposite to that shown.

Post-Deflection Acceleration (P.D.A.)

When an oscilloscope tube is used to display high-frequency waveforms or fast transients, the trace has to be brightened otherwise it is difficult to discern some of the detail. The trace brightness may be increased by increasing the beam current, i.e. reducing the grid bias voltage so that more electrons reach the screen phosphor. This, however, increases the spot size and so there is a limit to the maximum beam current in a c.r.t. The only other way is to increase the velocity of the electrons striking the screen by raising the accelerating voltage. If the voltages applied to the first and third anodes of the gun assembly in Fig. 7.5. were increased, the electron velocity would be increased but the deflection sensitivity of the deflector plates would be reduced since the electrons would spend less time in the deflecting fields. Higher deflecting voltages could be used but this is difficult at high frequencies.

The above problems are solved in a P.D.A. tube by increasing the velocity of the elctrons **after they have been deflected**. The basic idea is that if the final anode of the gun assembly is held at a voltage of about 1kV, the electrons will have relatively low velocity when passing through the deflector plates thus ensuring high deflection sensitivities. If the electrons are accelerated after deflection it will have little or no effect on deflection sensitivity. One method of achieving post deflection acceleration of the electrons is to use a high-resistance coating on the inside of the c.r.t. deposited in the form of a helix as shown in Fig. 7.10. The screen end of the helix coating is connected

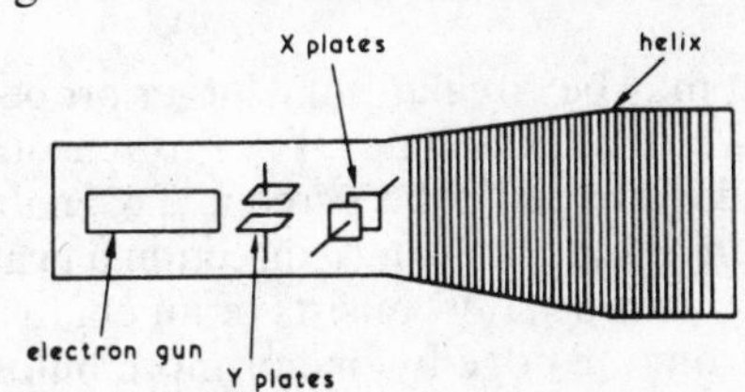

Fig. 7.10 Helix-type P.D.A. cathode ray tube.

to a high voltage (5 to 10kV) and the other end is connected to a potential at or near the final anode of the gun assembly. The electrostatic field set up by the helix accelerates the electrons leaving the deflection plates so that they strike the screen phosphor with high velocity. Due to the greater kinetic energy of the arriving electrons a larger light output from the spot is possible.

(3) VIEWING AREA

As mentioned at the beginning of this chapter, the screen is made of an electroluminescent material which fluoresces on being bombarded by the electron beam. The kinetic energy acquired by the electrons during motion is given up to the screen coating on impact causing it to fluoresce. Light is emitted from the screen phosphor in all directions at the point of impact as shown in Fig. 7.11(a). A large proportion of the light output is wasted being emitted backwards into the tube. This loss of light can be reduced by applying a thin layer of aluminium on one side of the screen phosphor as shown in Fig. 7.11(b). The aluminium layer acts as a highly reflective backing for the screen phosphor causing the rearward light to be reflected forwards and increasing the light output from the viewing side. There will be some small loss of electron energy in passing through the inter-atomic spaces of the aluminium and for this reason the coating is made very thin. 'Aluminising' the screen in this way also reduces the chance of screen 'burns'. The screen may be burnt by a stationary spot of light as the larger portion of the kinetic energy of the arriving electron beam causes heating of the screen phosphor; only about 10% is converted into light.

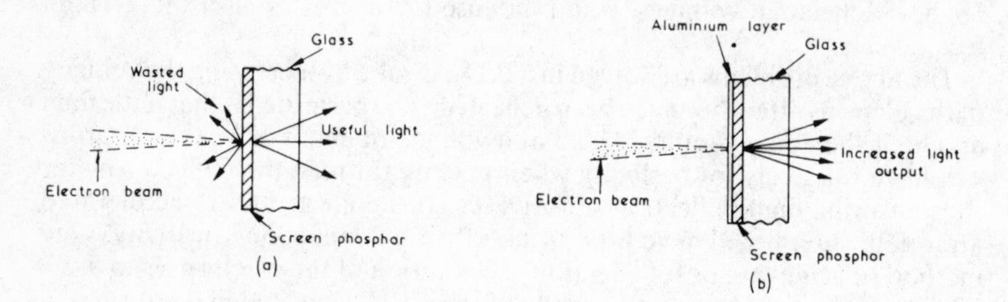

Fig. 7.11 Use of aluminising to improve light output from screen.

The face of the c.r.t may be circular but a larger proportion of the screen surface can be utilised if it is rectangular. For this reason, dual trace c.r.o.s often use a rectangular shaped tube screen. To enable the amount of deflection to be read in either the X or Y direction a **graticule** is fitted over the face of the screen. This usually consists of an engraved piece of plastic which is illuminated from the edge by one or more bulbs. Since the screen glass is thick and the phosphor coating is on the inside, parallex errors may arise when viewing the display. To avoid these errors the point to be read should be viewed at right angles.

C.R.T. SUPPLY VOLTAGES

Typical supply voltages for a basic oscilloscope tube are given in Fig. 7.12. An important point is that the mean potential of the X- and Y-plates must be approximately the same potential of the final (third) anode of the electron gun. It is therefore most convenient to arrange that the final anode is at earth potential and to place the cathode at a high negative potential of 1–2kV. This does not alter the principle of operation since the first anode (also at earth potential) is positive with respect to the cathode and provides the initial

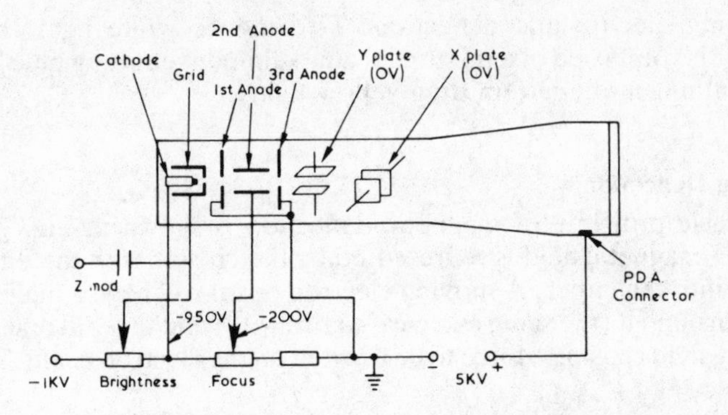

Fig. 7.12 Typical supply voltages for oscilloscope tube.

acceleration of the electrons on their way to the screen. The grid is held at a potential more negative than that of the cathode but its potential is made variable by the brightness control to vary the brilliance of the spot. Accurate focusing of the beam is achieved by providing a variable voltage to the second anode from the focus control. The X- and Y-plates are fed from their respective amplifiers in push-pull, i.e. one plate will change in voltage by $+V$ with respect to earth whilst the other changes by $-V$ with respect to earth, the mean voltage being zero as is required. A separate positive supply of 5–10kV is used to provide a post-deflection acceleration voltage for the tube.

TELEVISION DISPLAY TUBES

Cathode ray tubes used in television receivers operate similarly to those found in oscilloscopes. The main differences lie in the method of beam deflection, tube shape and operating voltages. The basic idea of a monochrome television c.r.t. is shown in Fig. 7.13. A large rectangular-faced c.r.t. is used, typical screen sizes being 22 in and 26 in (measured diagonally across the screen area). An electron gun assembly employing electrostatic focusing shoots a well-defined electron beam towards the screen phosphor at high velocity. Magnetic deflection of the beam is utilised as it is impracticable to obtain the large deflection angles required (of 90° or

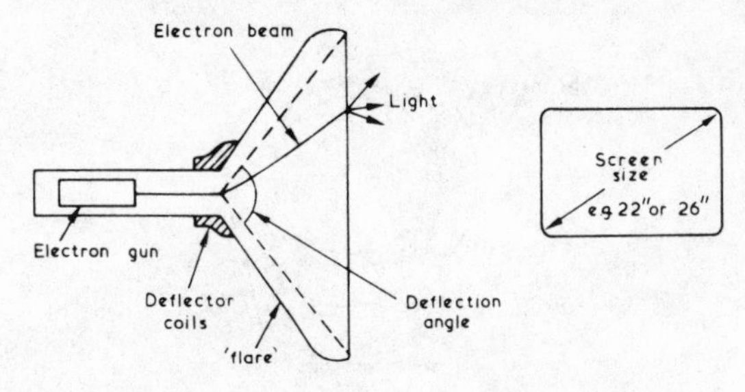

Fig. 7.13 Basic idea of monochrome television tube.

110°) using electrostatic deflection. To produce white light the screen phosphor is composed of a mixture of zinc sulphide (emitting blue light) and zinc cadmium sulphide (emitting yellow light).

Magnetic Deflection

The basic principle of magnetic deflection is shown in Fig. 7.14. The deflecting magnetic field is represented by the crosses with the direction of the field into the paper. A moving electron beam will have a magnetic field existing around it (the same as a current in a wire) and this will react with the deflecting field causing a force to be exerted on the electron beam. This force

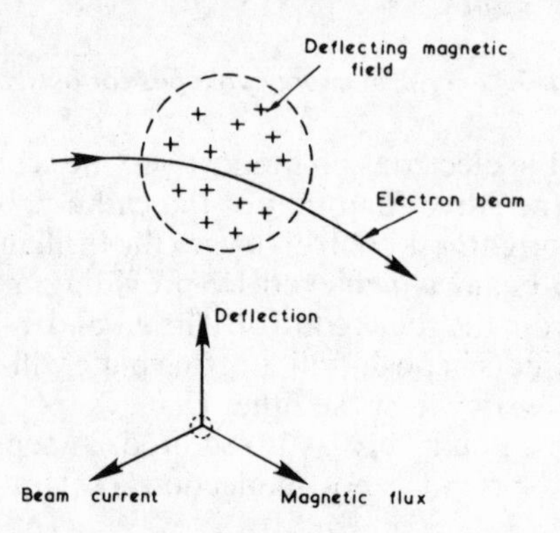

Fig. 7.14 Principle of magnetic deflection.

acts in a direction at right angles to both the direction of the deflecting field and the direction of the beam current thus deflecting the beam downwards as shown. While in the deflecting field the electrons travel in a curved path along the arc of a circle but on leaving the deflecting field they travel in a straight-line path.

To deflect the beam in both the horizontal and vertical directions two pairs of deflector coils are employed as shown in Fig. 7.15. Each pair is disposed at right angles to each other and both pairs are situated at the same place along the tube neck. This saves space and reduces the length of tube required.

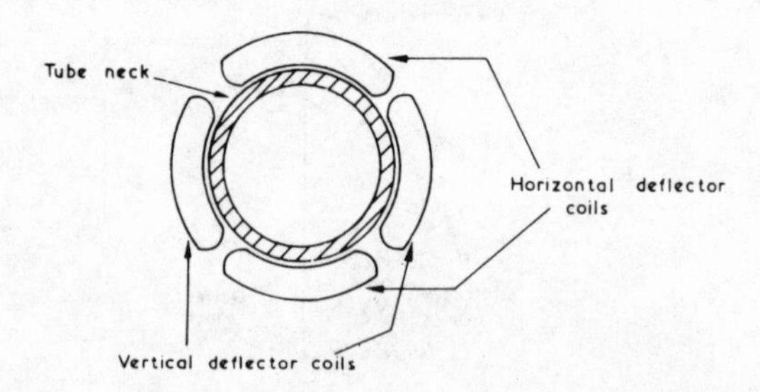

Fig. 7.15 Magnetic deflection in two directions at right-angles.

With electric deflection one set of plates follows the other, it not being practical to mount both sets of plates in the same position.

Deflection due to each pair of coils is considered separately in Fig. 7.16. The horizontally disposed coils (the field coils) of Fig. 7.16(a) produce deflection of the beam in the vertical direction. The direction of deflection

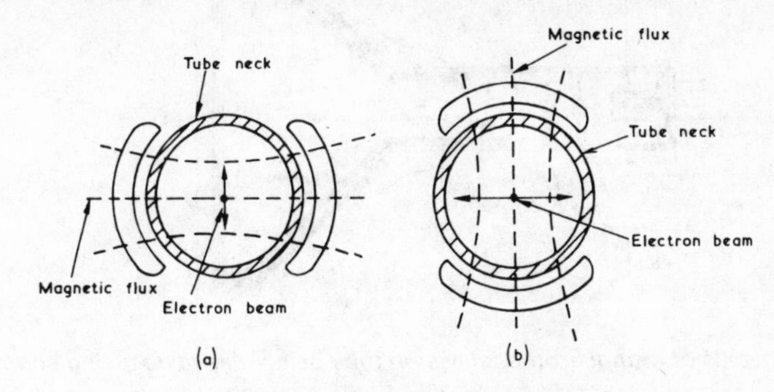

Fig. 7.16 Vertical and horizontal deflection of beam.

(up or down) may be altered by reversing the direction of the magnetic flux which is achieved by reversing the direction of the current flowing in the coils. Fleming's left-hand rule may be used to find the direction of deflection but remember that electron flow is in the opposite direction to conventional current flow. The vertically disposed coils (the line coils) of Fig. 7.16(b) produce deflection of the beam in the horizontal direction. The direction of deflection (left or right) may be altered by reversing the direction of current flow in the coils. The amount of deflection produced depends upon the strength of the magnetic field or the current in the deflector coils. To achieve a concentrated magnetic field passing through the tube neck, the coils are usually wound on a ferrite core and often the coils are taken up the flare of the tube to permit wide deflection angles.

Supplies to Monochrome Television C.R.T.

Further details showing the construction of a typical c.r.t. for use in a monochrome t.v. receiver are shown in Fig. 7.17. The electron gun assembly is composed of the heater, cathode, grid and four anodes. The first anode A1 is held at about 600–700V positive with respect to the cathode and provides the initial acceleration of the electrons on their way to the screen. The grid is placed at a negative potential with respect to cathode and by varying the grid voltage the brightness of the trace may be altered. Once the electrons leave the aperture in A1, they are accelerated to a high velocity as A2 is held at a potential of +15 to +20kV depending upon the tube size. Anodes A2, A3, and A4 form an electrostatic focusing lens with the focus of the beam controlled by adjustment of the potential applied to A3. The e.h.t. voltage required for A2 (also A4) is supplied via an e.h.t. connector on the tube flare, an internal conductive coating of graphite (called the 'aquadag') and

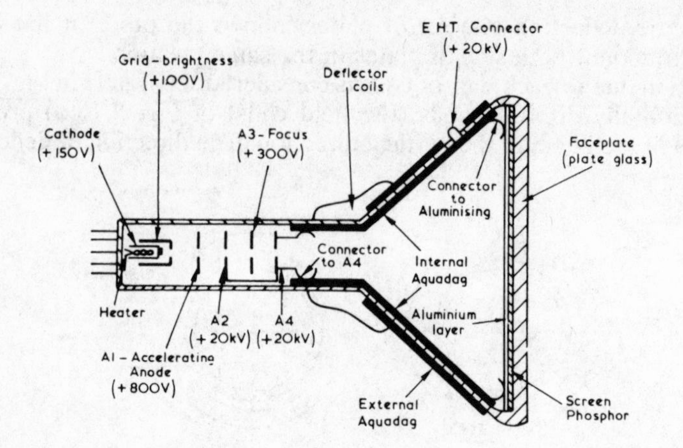

Fig. 7.17 Details of monochrome television tube using electric focusing and magnetic deflection.

spring clips as shown. At the back of the screen phosphor there is a thin layer of aluminium which is used to increase the light output and to reduce the possibility of burning the screen coating. The aluminium layer is held at the final anode potential and thus provides the return path for the electrons striking the screen. A further conductive coating applied over the external surface of the tube flare is also used which is normally connected to the chassis of the receiver via a spring clip. The external and internal coatings together with the interposing glass layer of the tube envelope form a small capacitor (500–1000pF) which is used to provide smoothing of the e.h.t. supply.

The line and field deflector coils are fed with sawtooth currents at 15 625Hz and 50Hz respectively, deflecting the beam over the screen to form the television raster.

To produce the television picture the electron beam is intensity-modulated by applying the video signal output of the receiver to the cathode of the c.r.t. This signal alters the grid-to-cathode bias and hence the intensity of the light output from the screen.

Colour Television C.R.T.

The basic principle of one type of colour display c.r.t. is shown in Fig. 7.18(a). Here three separate electron guns, arranged horizontally in-line, shoot fine electron beams towards the screen. The outer two guns are tilted slightly inwards so that the three beams cross over at a point lying in the plane of the shadow-mask as shown. The shadow-mask is made from thin sheet steel and has slots in it through which electrons from the three beams pass. After passing through the mask the beams fall on phosphor stripes deposited on the inside of the glass faceplate which emit light in the primary colours of red, green and blue. The electron beams are carefully aligned so

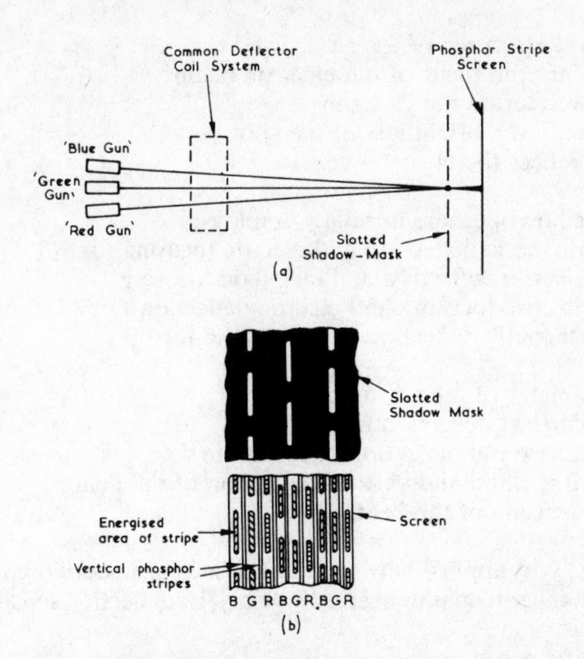

Fig. 7.18 Principle of in-line colour c.r.t.

that each beam falls on its respective colour phosphor stripe, thus energising small areas of the stripes as shown in Fig. 7.18(b). It is because the energised areas are so small that the eye does not see the individual coloured areas but only their additive mixture.

Each beam is electrostatically focused by anodes in its electron gun assembly but all three beams are deflected by a common deflection coil assembly using magnetic deflection. Apart from the need to provide three electron guns inside the neck of the c.r.t. and the special requirement of a shadow-mask, many of the constructional features of a colour c.r.t. are similar to those of a monochrome tube. A higher e.h.t. supply (up to 24kV) is required for the final anode and a higher focus voltage (about 5kV) is supplied to the focusing anode.

QUESTIONS ON CHAPTER SEVEN

(1) The 'emitting' material in a c.r.t. is:
- (a) Tungsten
- (b) Copper
- (c) Nickel
- (d) Barium and strontium oxides.

(2) The grid electrode of a c.r.t. is used to:
  (a) Vary the focus of the electron beam
  (b) Accelerate the electrons
  (c) Vary the brightness of the spot
  (d) Deflect the electron beam.

(3) An oscilloscope tube normally employs:
  (a) Magnetic deflection and electric focusing
  (b) Electric deflection and magnetic focusing
  (c) Electric focusing and electric deflection
  (d) Magnetic deflection and magnetic focusing.

(4) The X-plates of a c.r.t. produce:
  (a) Vertical deflection of the beam
  (b) Horizontal deflection of the beam
  (c) Horizontal and vertical deflection of the beam
  (d) Focusing of the beam.

(5) When 20V is applied between the Y-plates of an oscilloscope tube the spot on the screen is deflected by 4cm. The deflection sensitivity of the plates is:
  (a) 80V per cm
  (b) 5V per cm
  (c) 0·2V per cm
  (d) 2V per cm.

(6) The screen phosphor of a c.r.t. is 'aluminised' to:
  (a) Increase the light output
  (b) Increase the electron velocity
  (c) Provide a graticule
  (d) Reduce the spot size.

(7) A television tube normally employs:
  (a) Magnetic deflection and electric focusing
  (b) Electric deflection and magnetic focusing
  (c) Magnetic deflection and magnetic focusing
  (d) Electric deflection and electric focusing.

(Answers on page 180)

CHAPTER EIGHT

LCR CIRCUITS

THIS CHAPTER IS concerned with the effects of inductance, capacitance and resistance in d.c. and a.c. circuits, an understanding of which provides an important basis for the study of the diverse range of circuits used in electronic equipment.

L, *C* AND *R* IN D.C. CIRCUITS

CR Time-Constant

Consider the circuit of Fig. 8.1(a) where a capacitor C is to be charged via a resistor R to the d.c. supply of 10V. At the instant the switch is closed, the voltages in the circuit are as shown in Fig. 8.1(b). Initially all of the supply voltage appears across R, there being no voltage drop across C (capacitor uncharged). The initial charging current I that flows is given by Ohm's law,

$$I = \frac{E}{R} = \frac{10}{10} = 1\text{A}.$$

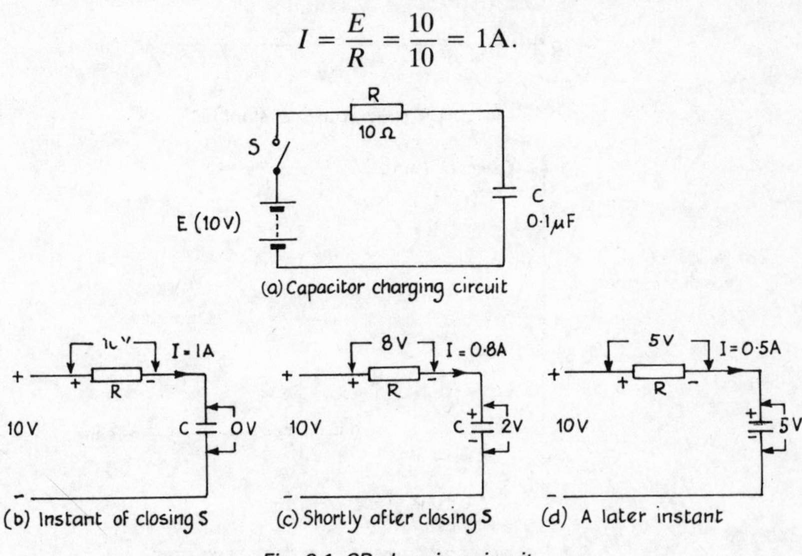

Fig. 8.1 CR charging circuit.

As the current commences to flow, the capacitor charges and the voltage across it rises. Suppose that shortly after closing the switch, the voltage across the capacitor has risen to 2V, see Fig. 8.1(c). Since the sum of the voltages across *C* and *R* must be equal to the supply voltage at all times, there will be only 8V across *R*. Accordingly, the charging current will now reduce to 8/10 = 0.8A. The capacitor voltage will continue to rise but due to the reduction in charging current, the rate of rise in voltage will not be so great as it was initially.

At a later instant, when, say, the voltage across the capacitor has risen to 5V, see Fig. 8.1(d), there will be only 5V across *R* and the charging current will reduce to 5/10 = 0.5A, reducing further the rate of rise in voltage across *C*.

This process continues with the voltage across *C* rising towards 10V and the voltage across *R* falling towards 0V. Because the charging current reduces as the capacitor voltage rises, the rise of voltage follows a non-linear (exponential) relationship with time, as illustrated in Fig. 8.2(a). The decaying charging current is shown in Fig. 8.2(b). Eventually, when the

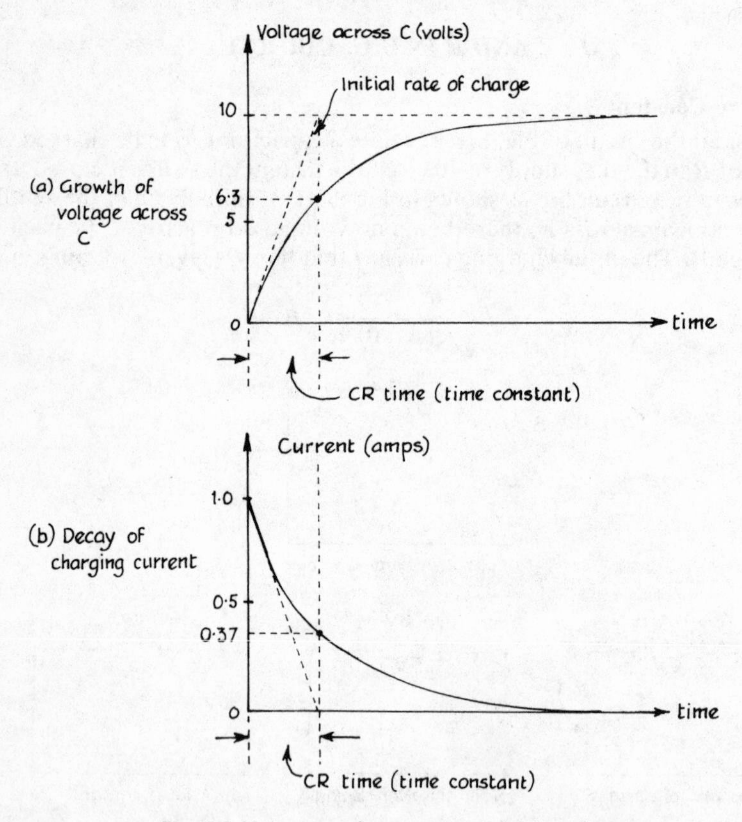

Fig. 8.2 Voltage and current as capacitor charges.

voltage across C reaches 10V, the voltage across R will be 0V (no current will flow) and the capacitor will be fully charged to the supply voltage.

The **time constant** (or CR time) of the CR circuit is defined as the time (seconds) taken for the voltage across the capacitor to reach 0·63 of the applied voltage and is given by the product of C (farads) and R (ohms). For the values given in Fig. 8.1(a),

$$\text{Time constant} = 0{\cdot}1 \times 10^{-6} \times 10 \text{ seconds} = 1\mu\text{s}$$

Time constant may alternatively be defined as the time taken for the voltage across the capacitor to reach the supply voltage if the initial rate of charge had been maintained. It will be seen that the voltage across C gradually approaches the supply voltage once it has exceeded the 0·63 × supply voltage point. For practical purposes it may be considered that the capacitor is fully charged after an interval corresponding to **5 × time constant**.

Example 1

A capacitor of 10nF is charged via a resistance of 100kΩ to a d.c. supply voltage of 50V. Calculate:

(a) the time constant of the circuit;
(b) the initial charging current at switch-on;
(c) the voltage across the capacitor after a period equal to the time constant;
(d) the approximate time for the voltage across the capacitor to reach 50V.

Solution

(a)
$$\begin{aligned}\text{Time constant} &= C \times R \text{ seconds}\\ &= 10 \times 10^{-9} \times 10^{5} \text{ seconds}\\ &= 10 \times 10^{-9} \times 10^{5} \times 10^{3} \text{ ms}\\ &= 1\text{ms}\end{aligned}$$

(b)
$$\text{Initial charging current} = \frac{E}{R} = \frac{50}{10^5}\text{A} = 0{\cdot}5\text{mA}$$

(c) After a period equal to 1ms,
$$\begin{aligned}\text{Voltage across the capacitor} &= 0{\cdot}63 \times 50\text{V}\\ &= 31{\cdot}5\text{V}\end{aligned}$$

(d)
$$\begin{aligned}\text{Time taken for capacitor to charge fully} &= 5 \times \text{times constant}\\ &= 5\text{ms}\end{aligned}$$

Discharging Circuit

Consider now the circuit of Fig. 8.3(a), where the capacitor which was previously charged to 10V is allowed to discharge through R. At the commencement of discharge,

$$\text{Current } I \text{ flowing} = \frac{V}{R} = \frac{10}{10} = 1\text{A}$$

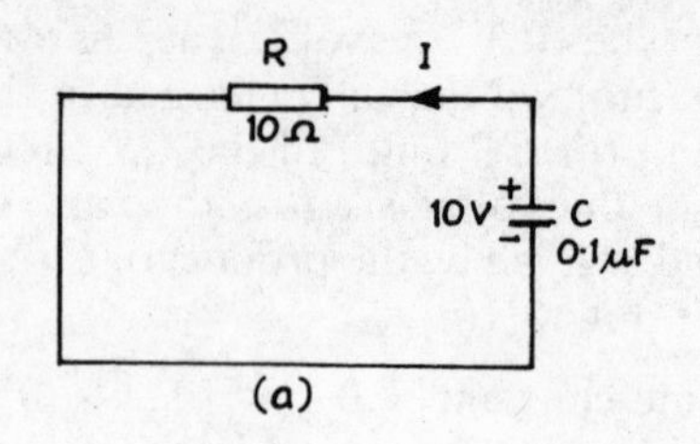

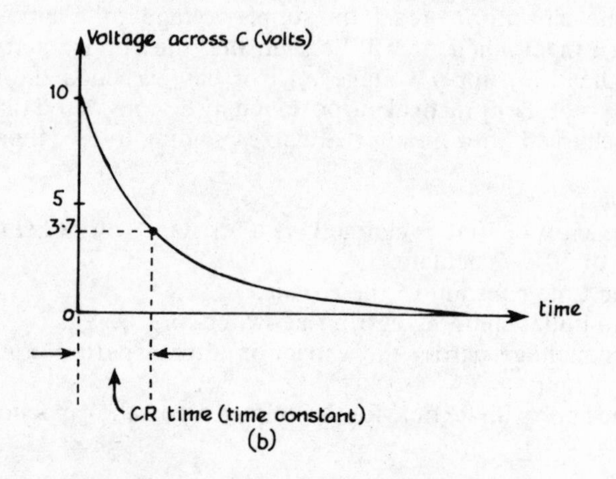

Fig. 8.3 Voltage as capacitor discharges.

When current flows in R, charge is lost from C, resulting in the voltage across C falling. This fall in voltage will reduce the magnitude of the current flowing, thus the rate at which the capacitor voltage falls is not constant but follows the non-linear (exponential) form shown in Fig. 8.3(b). The decay of current with time will be exactly the same as in Fig. 8.2(b).

The time constant of the discharge circuit may be defined as the time taken for the capacitor voltage to fall to 0·37 of its initial value. The time constant is, as before, given by C (farads) $\times$ R (ohms) seconds and is precisely the same for the charging circuit since the same value of resistor is used. After a period equal to approximately 5 $\times$ time constant, the capacitor may be considered to be completely discharged (0V).

Example 2

A capacitor of value 3μF previously charged to 80V is discharged through a 1MΩ resistor. Calculate:

(a) the time constant;
(b) the initial discharge current;
(c) the voltage across the capacitor after a period equal to the time constant;
(d) the approximate time for the capacitor to discharge fully.

Solution

(a) Time constant $= C \times R$ seconds
$= 3 \times 10^{-6} \times 10^{6}$ seconds
$= 3s$

(b) Initial discharge current $= \frac{V}{R} = \frac{80}{10^{6}}A = 80\mu A$

(c) After a period equal to 3s the voltage across the capacitor will be $0{\cdot}37 \times 80 = 29{\cdot}6V$

(d) Time taken for capacitor to discharge fully $= 5 \times$ time constant
$= 15s$

The diagram of Fig. 8.4 shows the effect of varying the time constant when the capacitor is charging. Clearly, the longer the time constant the longer will be the time required for the capacitor to charge fully or to reach any specified voltage level.

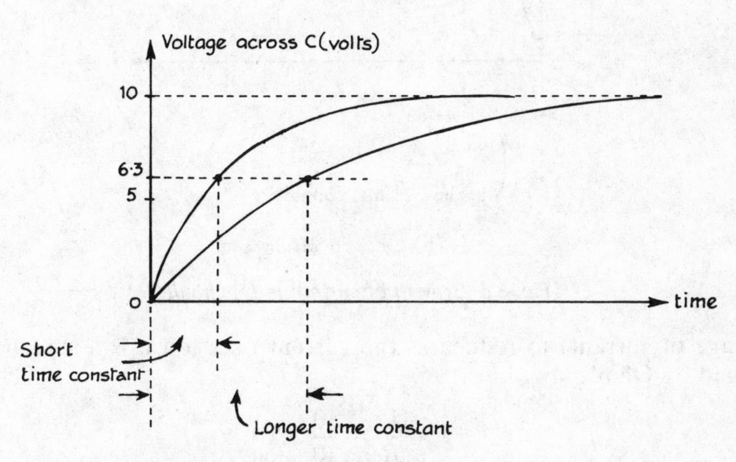

Fig. 8.4 Effect of varying time constant (CR time).

LR Time Constant

Consider the circuit of Fig. 8.5(a) consisting of an inductor L in series will a resistance R (which may be just the resistance of the inductor). If the switch is closed, current will commence to flow but is prevented from building-up rapidly since an e.m.f. will be induced in the inductor. The back e.m.f. acts in opposition around the circuit to the applied voltage E thus reducing the rise of current. As a result, the current rises non-linearly as shown in Fig. 8.5(b). It will be seen that the rate of change of current reduces as the current rises thus causing the back e.m.f. (which is proportional to the rate of

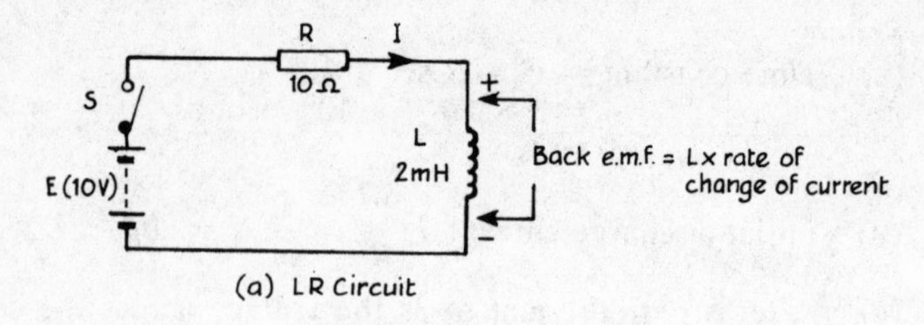

(a) LR Circuit

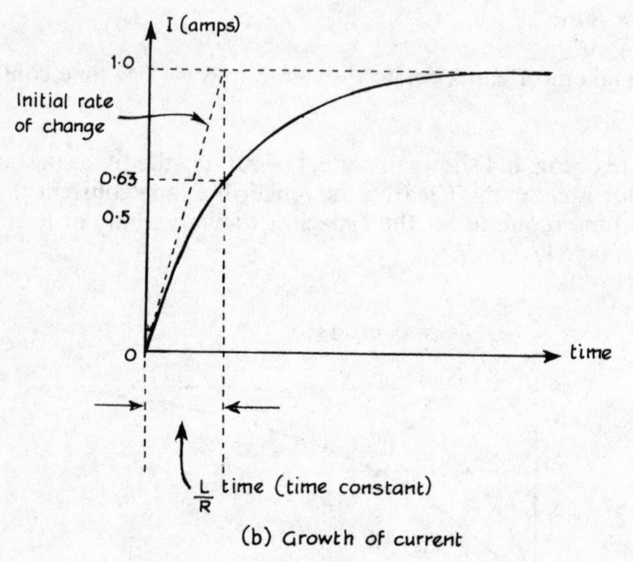

(b) Growth of current

Fig. 8.5 Growth of current in LR circuit.

change of current) to reduce as the current rises towards a steady value settled by Ohm's law, i.e.

$$I = \frac{V}{R} = \frac{10}{10}\ 1\text{A}$$

The time constant of the *LR* circuit may be defined as the time taken for the current to reach 0·63 of its final steady value and is equal to *L*/*R* seconds, where *L* is in henrys and *R* is in ohms. For the values given,

$$\text{Time constant} = \frac{L}{R}\ \text{seconds}$$

$$= \frac{2 \times 10^{-3}}{10}\ \text{seconds}$$

$$= 0{\cdot}2\text{ms}$$

For practical purposes the current will reach its steady level after a period equal to 5 × time constant.

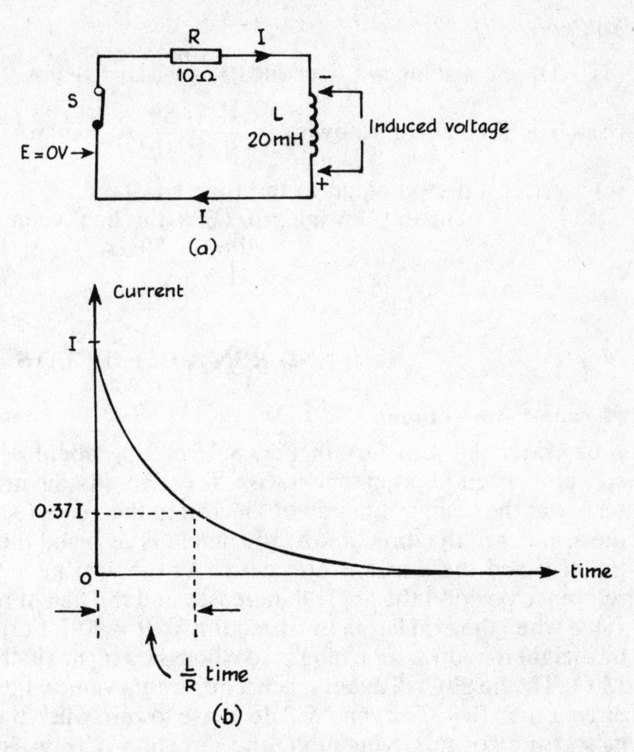

Fig. 8.6 Decay of current in an LR circuit.

If the applied voltage E is suddenly reduced to zero but the circuit is not broken, see Fig. 8.6(a), the current will start to decrease but an e.m.f. will be induced in L tending to keep the current flowing and in the same direction. The current will decrease as in Fig. 8.6(b) and will fall to 0·37 of its initial value (E/R) in a time equal to the time constant of L/R seconds. Thus an inductor tends to prevent changes of current through it, a property which is used in electronic circuits.

If an attempt is made to break an inductive circuit rapidly, a high voltage is induced in the inductor (since the induced e.m.f is proportional to the rate of change of current) which may be dangerous or damage the insulation between the turns of the inductor. This principle is put to use in car ignition coils, the starting of fluorescent lamps and the generation of flyback e.h.t. in television receivers.

Example 3

An inductor of 0·5H is connected in series with a resistor of 1kΩ across a 50V supply. Determine:

(a) the time constant of the circuit;
(b) the final value of current in the circuit;
(c) the current in the circuit after a period equal to the time constant.

Solution

(a) Time constant $= \frac{L}{R}$ seconds $= \frac{0{\cdot}5}{10^3}$ s $= 0{\cdot}5$ms

(b) The final current flowing $= \frac{E}{R} = \frac{50}{10^3}$ A $= 50$mA

(c) After a period equal to the time constant,

$$\begin{aligned}\text{Current flowing} &= 0{\cdot}63 \times \text{the final value}\\ &= 0{\cdot}63 \times 50\text{mA}\\ &= 31{\cdot}5\text{mA}\end{aligned}$$

L, *C* AND *R* IN A.C. CIRCUITS

Phasor Representation

Consider the arm OA of Fig. 8.7 rotating about a fixed point O at a constant speed in an anticlockwise direction. As the arm rotates it will be seen that the height of the tip of the arm to the x-axis (X – X′) varies. When the arm lies in the direction X, the height is zero and the angle (θ) between the arm and the x-axis is also zero. As the arm moves from the starting reference position the angle θ increases and the height reaches a maximum value when the arm lies in the direction Y (θ = 90°). Further rotation causes the height to reduce, reaching zero when the arm lies in the X′ direction (θ = 180°). The height will again reach a maximum value when the arm lies in the direction Y′ (θ = 270°) and will decrease to zero when the arm has moved to its starting position lying along the direction X (θ = 360°).

If the height of the arm from the x-axis is plotted against angle, a sinewave is produced as shown. The angle may be given in degrees or radians where 360° = 2π radians (1 radian = 57·296°).

The arm OA is the **phasor representation** of the sinewave and its length is equal to the **maximum** or **peak value** of the sinewave. The angular position of the **phasor** OA at any instant (measured from the reference direction OX) gives the angle traced out by the sinewave. After one revolution of the phasor corresponding to one cycle of the sinewave, OA will have moved

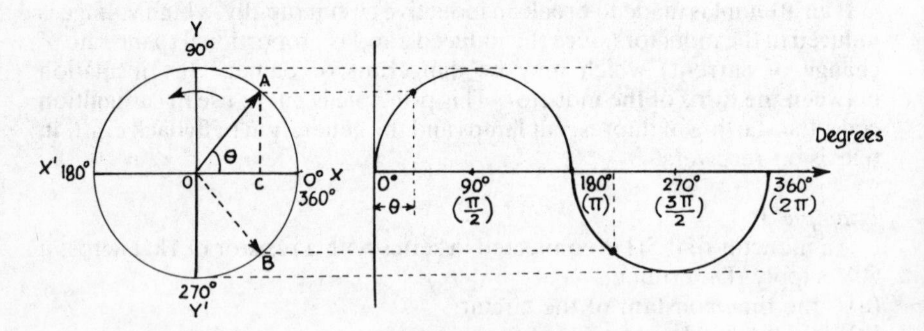

Fig. 8.7 Phasor representation of sinewave.

through an angle of 2π radians. If the sinewave has a frequency of f cycles per second (Hz), the phasor will move with an angular velocity of $2\pi f$ rad/s or ω rad/s (where $\omega = 2\pi f$).

Example 4

A sinewave has a frequency of 100Hz. Determine its angular velocity (angle covered every second).

Solution

$$\begin{aligned}\text{Angular velocity} &= 2\pi f \text{ rad/s} \\ &= 2 \times \pi \times 100 \text{ rad/s} \\ &= 628{\cdot}32 \text{ rad/s}\end{aligned}$$

Phase Difference

Voltages or currents found in d.c. circuits act either in the same direction or in opposite directions, but in a.c. circuits voltages or current can occur with various timing (phase) differences between their maximum peak values.

Consider Fig. 8.8(a) which shows two sinewaves $v1$ and $v2$ of different amplitudes and of the same frequency but with their peak values occurring at

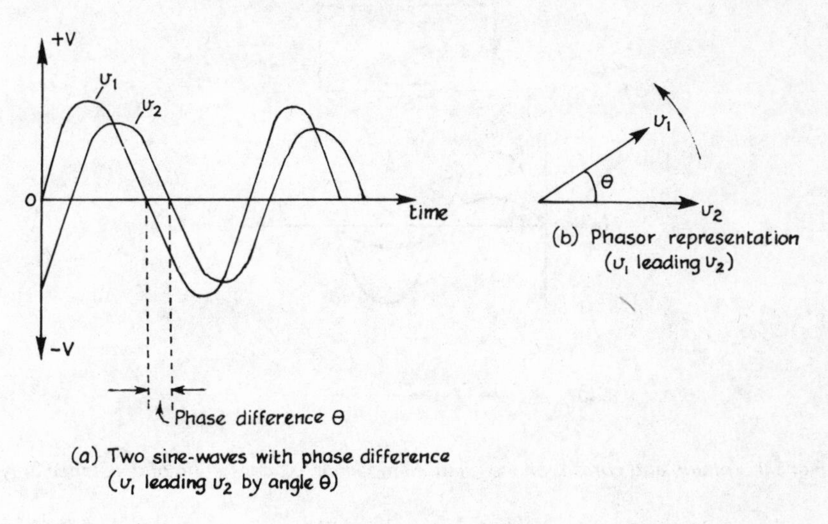

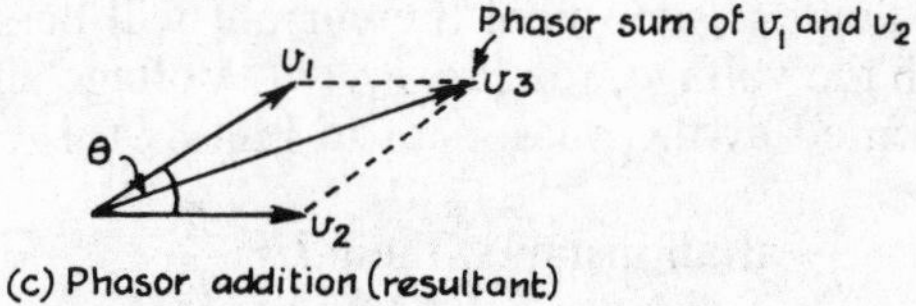

Fig. 8.8 Phase difference.

different time instants. The two sinewaves are said to have a phase difference with $v1$ leading $v2$ by an angle θ; although the x-axis is given in time it is related to the angle of the wave since the time period of one cycle represents 2π radian.

The sinewaves may be represented by the phasor diagram of Fig. 8.8(b) where the phasors $v1$ and $v2$ each have a length proportional to the peak amplitudes of the sinewave voltages and are shown with a phase difference of θ degrees. There is no need to show the phasors rotating since both sinewaves are of the same frequency and the phase difference between them will be constant. If the phasors $v1$ and $v2$ are added together, the result will be as in Fig. 8.8(c) where the resultant $v3$ is another sinewave having the same frequency but a different amplitude from $v1$ or $v2$. It will be seen later in this chapter the usefulness of phasor representation of sinewaves and the importance of phase difference, which is the cause of many interesting phenomena that occur when alternating current flows in electrical circuits.

Circuit With Resistance Only

If a voltage E is applied to the circuit of Fig. 8.9(a) then at any instant when the voltage is e, the current i is given by $i = e/R$. Thus the current flowing in the circuit is proportional to the voltage at any instant, and if the

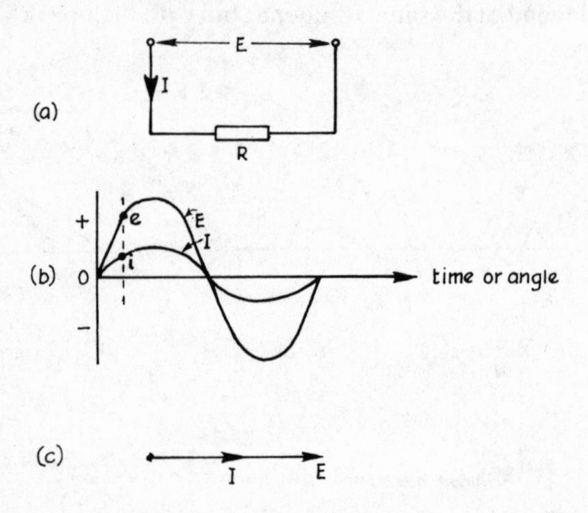

Fig. 8.9 Voltage and current phase relationship in circuit consisting of resistance only.

applied voltage is sinusoidal the current will be sinusoidal and will be in phase with the voltage, see Fig. 8.9(b). Voltage and current may therefore be represented by the phasors as in Fig. 8.9(c).

Since $i = \dfrac{e}{R}$ at all instants, then $I = \dfrac{E}{R}$

where I and E are the r.m.s. values. If the frequency of the applied voltage is changed and the r.m.s. value of the voltage remains constant, the r.m.s. value of the current remains constant. Thus the value of R does not vary with frequency.

Summarising, it may be said that resistance in an a.c. circuit behaves exactly the same as it would in a d.c. circuit.

Circuit with Inductance Only

Consider the circuit of Fig. 8.10(a) consisting of an inductance L henrys and assume that the resistance of the inductor is negligible. Suppose that a sinusoidal current I is passed through the inductor. Since the current is continually changing there will be a voltage induced in the inductor which is proportional to the rate of change of current. As the rate of change of current is greatest as the current passes through the zero datum line, i.e. at points A, C, E, etc., the induced voltage will be at a maximum. At points B, D, etc. where the current is constant, there will be no induced voltage.

The direction or polarity of the induced voltage will be such that it opposes the change which is causing the voltage (Lenz's Law). At instant A it is the rise of current that is causing the induced voltage thus the voltage will be in such a direction as to oppose the current rise, i.e. it will be negative. As we move towards point B the rate of rise of current decreases hence the induced voltage becomes less, until at B where the current is steady for a brief instant

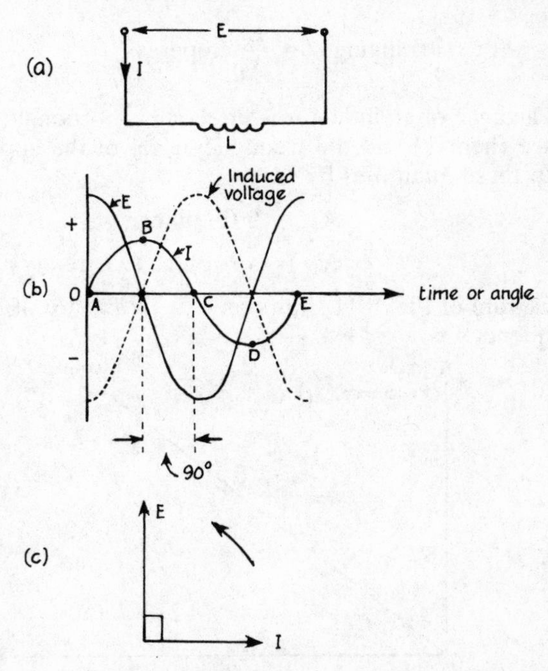

Fig. 8.10 Voltage and current phase relationship in circuit consisting of inductance only.

and the induced voltage is zero. After point B, the current starts to decrease so that the induced voltage will reverse direction so as to try to maintain the flow of current. The rate of decrease of current reaches a maximum at C and hence the induced voltage will be a maximum at this point. From C to D the rise of current is in the opposite direction and the induced voltage is positive, opposing the current rise in the negative direction. Finally, from D to E the current is decreasing and the induced voltage is in a direction such as to assist the flow of current.

Thus, if we are to force a current through the inductor, the applied voltage E must be equal and opposite to the induced voltage as shown in Fig. 8.10(b). It will be seen that the **applied voltage E leads the current I by 90°** or looking at it the other way, the current in an inductor lags the applied voltage by 90°. We may therefore represent the voltage and current in an inductive circuit by the phasors of Fig. 8.10(c).

Inductive Reactance

In an inductive circuit there is a certain ratio between the voltage and the current and this ratio is called the **inductive reactance** (X_L). It is a measure (in ohms) of the opposition offered by the inductor to the flow of alternating current and is given by

$$X_L = \frac{E}{I} \text{ ohms}$$

$$\text{or rearranging, } I = \frac{E}{X_L} \text{ amperes}$$

The reactance of an inductor is directly proportional to the value of the inductance (henrys) and the frequency (Hz) of the applied e.m.f. and is related to these quantities by

$$X_L = 2\pi f L \text{ ohms}$$

$$\text{or } X_L = \omega L \text{ ohms (where } \omega = 2\pi f).$$

The diagram of Fig. 8.11 shows how the reactance of an inductor varies with frequency.

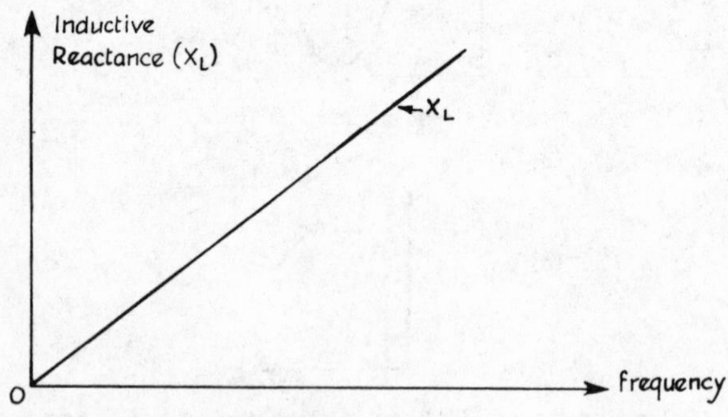

Fig. 8.11 Variation of inductive reactance with frequency.

Example 5

An alternating voltage of 10V r.m.s. at a frequency of 2kHz is applied to an inductor of value 5mH. Determine:

(a) the reactance of the inductor;
(b) the current flowing in the inductor.

Solution

(a)

$$X_L = 2\pi fL \text{ ohms}$$
$$= 2 \times \pi \times 2 \times 10^3 \times 5 \times 10^{-3}$$
$$= 62{\cdot}83 \text{ ohms}$$

(b)

$$I = \frac{E}{X_L} \text{ amperes}$$
$$= \frac{10}{62{\cdot}83} \text{ A}$$
$$= 159{\cdot}15\text{mA (r.m.s.)}$$

Circuit With Capacitance Only

We will now consider a circuit containing capacitance only as in Fig. 8.12(a). If d.c. were applied to the circuit then no current would flow except for an initial charging current. This is not the case if a.c. is applied to the circuit and is due to the fact that whenever the voltage across a capacitor is changed a charge flows in or out of it, and thus a current must flow. With a.c. the voltage is changing continuously, hence a resulting current flows in the circuit.

Suppose a sinusoidal voltage E is applied to the circuit as in Fig. 8.12(b), then the current that flows is proportional to the rate of change of voltage ($I = C \times$ rate of change of voltage). This is a maximum at points A, C and E

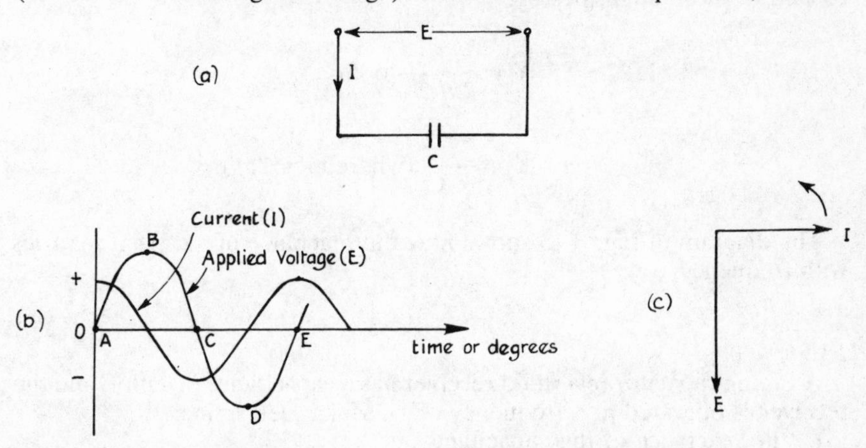

Fig. 8.12 Voltage and current phase relationship in circuit containing capacitance only.

but zero at points B and D. Thus, as the voltage rises rapidly at A a large current will flow causing the capacitor to charge up but as the rate of rise of voltage decreases towards B the current decreases. At B when the voltage is steady for a brief instant the current will be zero. Between B and C the voltage decreases causing the capacitor to discharge and for current to flow in the opposite direction reaching a maximum at C where the rate of change of voltage is at a maximum. Between C and D the capacitor is charged in the opposite direction and the current flows in the same direction as the voltage, but reaches zero at D where the rate of change of voltage is zero. Between D and E the capacitor again discharges.

It may thus be considered that a current flows in and out of the capacitor for each half cycle of the applied voltage. As can be seen from Fig. 8.12(b), the **current leads the voltage by 90°** and this may be represented by the phasors of Fig. 8.12(c).

Capacitive Reactance

In a capacitive circuit there is a certain ratio between the voltage and the current and this ratio is known as the capacitive reactance (X_c). It is a measure (in ohms) of the opposition offered by the capacitor to the flow of alternating current and is given by:

$$X_c = \frac{E}{I} \text{ ohms}$$

$$\text{or rearranging, } I = \frac{E}{X_c} \text{ amperes}$$

The reactance of a capacitor is inversely proportional to the value of the capacitor (farads) and to the frequency (Hz) of the applied voltage and is related to these quantities by:

$$X = \frac{1}{2\pi f C} \text{ ohms}$$

$$\text{or } X = \frac{1}{\omega C} \text{ (where } \omega = 2\pi f\text{).}$$

The diagram of Fig. 8.13 shows how the reactance of a capacitor varies with frequency.

Example 6

A tuning capacitor in a radio receiver has a capacitance of 250pF and the receiver is operated at a frequency of 1·2MHz. Determine:

(a) the reactance of the capacitance;

(b) the current in the capacitor if there is 10V across it.

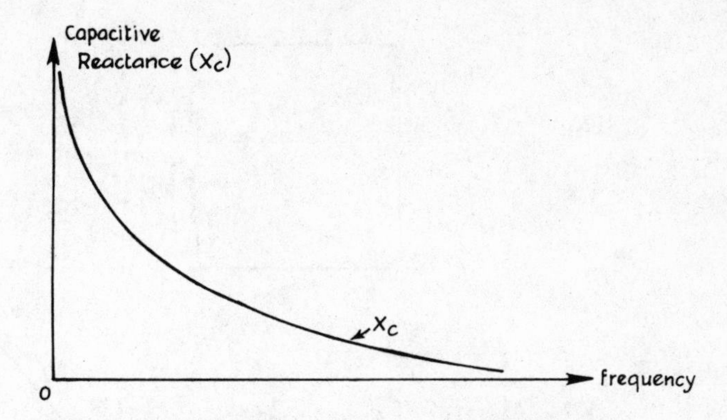

Fig. 8.13 Variation of capacitive reactance with frequency.

Solution

(a)

$$X_c = \frac{1}{2\pi f C} \text{ ohms}$$

$$= \frac{1}{2 \times \pi \times 1{\cdot}2 \times 10^6 \times 250 \times 10^{-12}} \text{ ohms}$$

$$= 530{\cdot}5 \text{ ohms}$$

(b)

$$I = \frac{E}{X} \text{ amperes}$$

$$= \frac{10}{530{\cdot}5} \text{ A}$$

$$= 18{\cdot}85\text{mA}$$

Circuits with *L* and *R* in series

In the previous sections we considered 'pure inductance' and 'pure capacitance' which are virtually impossible to achieve in practice and were dealt with simply to provide an understanding of the practical circuits found in electronics. For example, an inductor will always possess some d.c. resistance thus a practical inductor may be represented by an inductance in series with a resistance as in Fig. 8.14(a).

Since L and R are in series the same current I must flow through both components. There will be a voltage across R (V_R) equal to $I \times R$ and in phase with the current, also a voltage across L (V_L) equal to $I \times X_L$ but leading the current by 90°. In a d.c. circuit, to find the magnitude of the supply voltage (V) we would simply add the two voltages together. This does

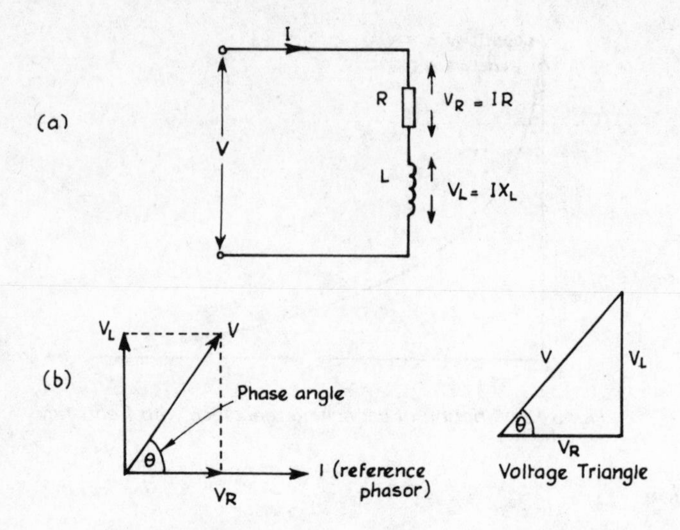

Fig. 8.14 L and R in series.

not solve the problem when the supply is a.c. since the two voltages have a different phase relationship to one another. These phase relationships may be taken into account by drawing the phasors for the circuit, see Fig. 8.14(b).

As the current is the common factor it can be used as the reference phasor. The voltage across the resistor may be represented by a phasor in phase with the current and the voltage across the inductor by a phasor leading the current by 90°. The supply voltage V is given by the **phasor sum** as shown and is seen to **lead the current by the phase angle** θ. From the phasor diagram, a **voltage triangle** may be extracted where the sides have lengths proportional to V_R, V_L and V.

Example 7

Consider the diagram of Fig. 8.15(a) where the voltages across R and L are 3V and 4V respectively. Phasors may be drawn to scale for V_R and V_L and the resultant of the phasor addition measured to find the supply voltage V, as in Fig. 8.15(b). The phase angle θ may be measured.

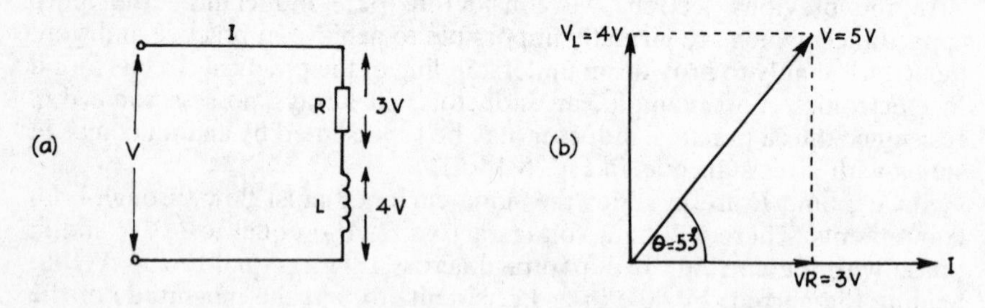

Fig. 8.15 Phasor addition of voltages across L and R.

The applied voltage V may be calculated from:

$$V = \sqrt{(V_R^2 + V_L^2)}$$

Using the values given in Fig. 8.15(a) gives:

$$V = \sqrt{(3^2 + 4^2)}$$
$$= \sqrt{(9 + 16)}$$
$$= \sqrt{25}$$
$$= 5\text{V}$$

Impedance

The total opposition to the flow of current of the circuit of Fig. 8.16(a) is made up of the resistance R which is constant regardless of frequency and the inductive reactance of L which varies as the frequency varies. This opposition to current flow is now given a new name **impedance** (Z) and is equal to the ratio of the applied voltage V to the current flowing I, i.e.

$$Z = \frac{V}{I} \text{ ohms}$$

If the values of R and X_L (in ohms) are known we do not simply add the values together to find the circuit impedance, since the resistance and reactance have a different phase relationship relative to the current flowing. The circuit impedance may be found by drawing phasors as in Fig. 8.16(b)

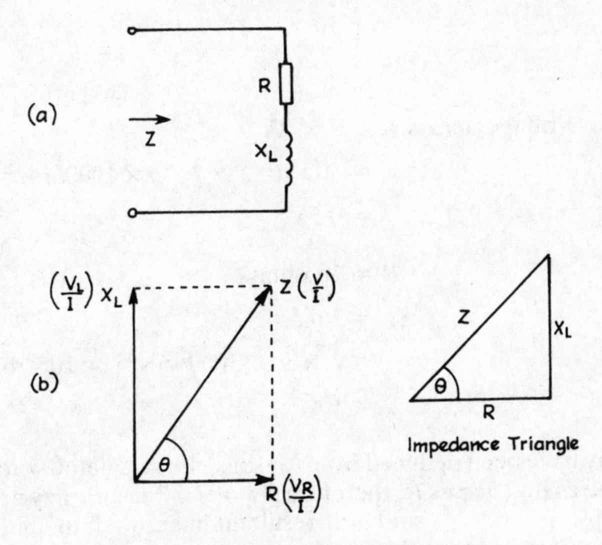

Fig. 8.16 Impedance of L and R in series.

and is the simple process of dividing V_R, V_L and V by the common current component I. Thus the impedance Z is the **phasor addition** of R and X_L and the voltage triangle can be redrawn as the **impedance triangle**.

When the values of R and X_L are known, phasors for R and X_L may be drawn to scale and the resultant of the phasor addition measured to find the circuit impedance. Alternatively, the impedance may be calculated from:

$$Z = \sqrt{(R^2 + X_L^2)}$$

Example 8

A current of 50mA at a frequency of $1000/\pi$ Hz is passed through the series circuit of Fig. 8.17. Determine:

(a) the voltages across R and L;

(b) using phasor diagrams, the total impedance of the circuit and the voltage V across the circuit.

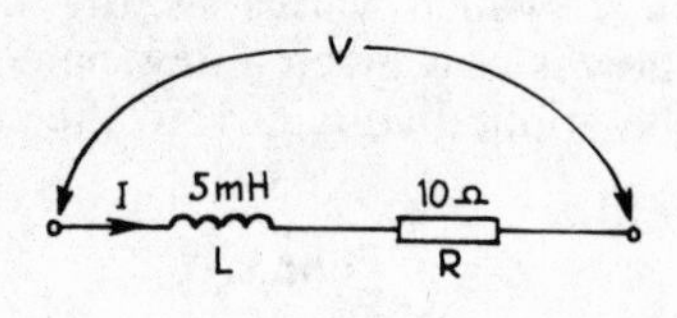

Fig. 8.17.

Solution

(a)

$$\text{Voltage across } R = I \times R$$
$$= 50 \times 10^{-3} \times 10 \text{ volts}$$
$$= 0{\cdot}5\text{V}$$

$$\text{Voltage across } L = I \times X_L$$
$$= 50 \times 10^{-3} \times 2 \times \pi \times 1000/\pi \times 5 \times 10^{-3} \text{ volts}$$
$$= 0{\cdot}5\text{V}$$

(b)

$$R = 10 \text{ ohms}$$
$$X_L = 2\pi f L$$
$$= 2 \times \pi \times 1000/\pi \times 5 \times 10^{-3} \text{ ohms}$$
$$= 10\Omega$$

(X_L may have been deduced from (a) since the calculated voltage across L is the same as that across R, therefore $X_L = R$.) Phasors may now be drawn to scale for R and X_L and the resultant measured to find the circuit impedance Z as in Fig. 8.18.

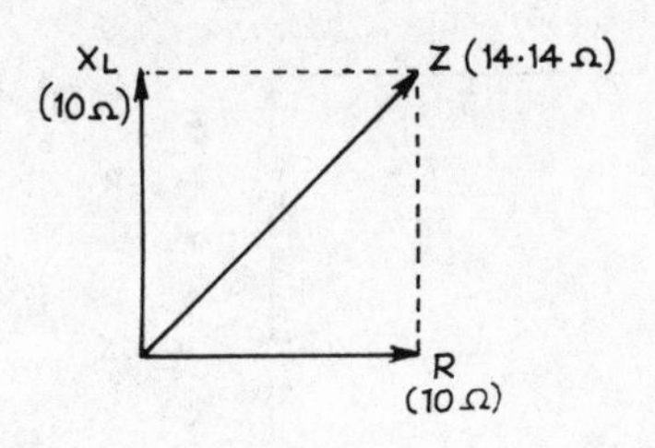

Fig. 8.18.

To find the applied voltage, phasors for V_R and V_L may be drawn as in Fig. 8.19 and the resultant measured. The above may be checked from:

$$I = \frac{V}{Z} = \frac{0{\cdot}707}{14.14}\text{ A} = 50\text{mA (given).}$$

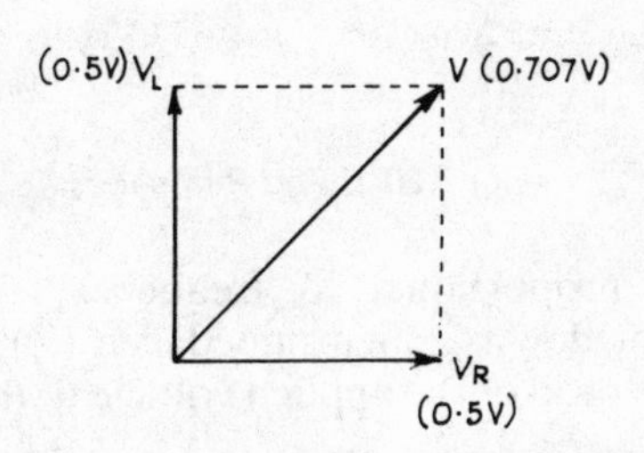

Fig. 8.19.

Circuits with *C* and *R* in series

Consider now a circuit consisting of a capacitor and a resistor in series, see Fig. 8.20(a). Since the same current I flows in both components, this is drawn as the reference phasor, Fig. 8.20(b). There will be a voltage across R equal to $I \times R$ and in phase with the current, also a voltage across C of magnitude $I \times X_c$ but lagging on the current by 90°. The supply voltage V is given by the phasor sum of V_R and V_C and is seen to be **lagging the current by the phase angle** θ. Again, a voltage triangle may be extracted from the phasor diagram where the sides have lengths proportional to V_R, V_C and V.

When the voltage across R and C are known, phasors for V_R and V_C may be drawn to scale and the resultant measured to find V. Alternatively, the supply voltage may be calculated from :

$$V = \sqrt{(V_R^2 + V_C^2)}$$

The total opposition to the flow of current is made up of the resistance R which is constant regardless of frequency, and the capacitive reactance of C

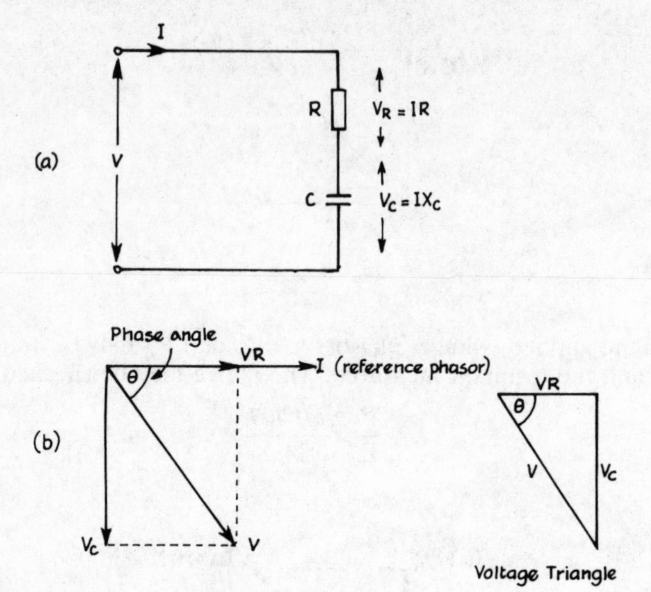

Fig. 8.20 C and R in series.

which is inversely proportional to frequency, see Fig 8.21(a). This opposition to current flow is called impedance (Z), as for the series *L*/*R* circuit is equal to the ratio of the applied voltage to the current flowing, i.e.

$$Z = \frac{V}{I} \text{ ohms}$$

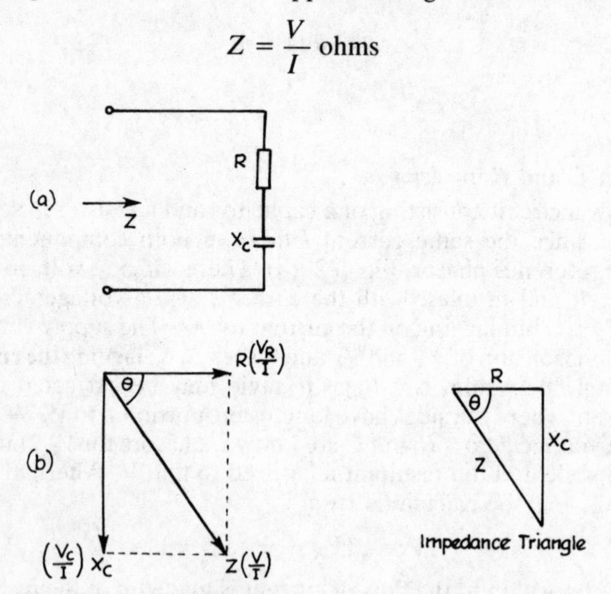

Fig. 8.21 Impedance of C and R in series.

To find the circuit impedance we cannot simply add R to X_C since the resistance and reactance have a different phase relationship to the current flowing. Thus once again, Z is found from the phasor addition of R and X_C, see Fig. 8,21(b). When the values of R and X_C are known, phasors for R and X_C may be drawn to scale and the resultant measured to find Z. Alternatively, Z may be calculated from:

$$Z = \sqrt{(R^2 + X_C^2)}$$

Example 9

If a current I of 100μA at a frequency of $2000/\pi$Hz flows in the circuit of Fig. 8.22, determine, using phasor diagrams:

(a) the impedance of the circuit;
(b) the applied voltage V.

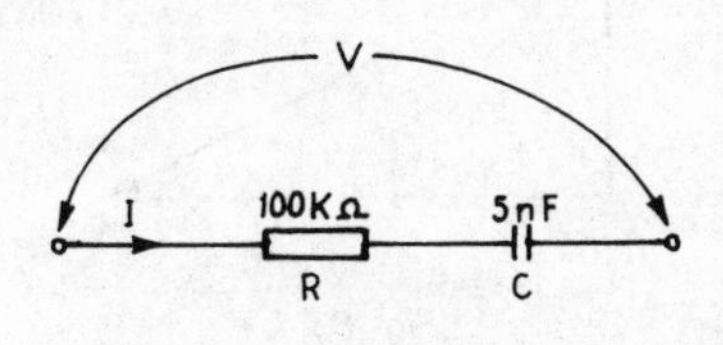

Fig. 8.22.

Solution

(a)

$$R = 100\text{k}\Omega \text{ (given)}$$

$$X_C = \frac{1}{2\pi f C}$$

$$= \frac{1}{2\pi \times \frac{2000}{\pi} \times 5 \times 10^{-9}} \text{ ohm}$$

$$= 50\text{k}\Omega$$

Phasors for R and X_C may now be drawn to scale as in Fig. 8.23 and the resultant measured to find Z (112kΩ).

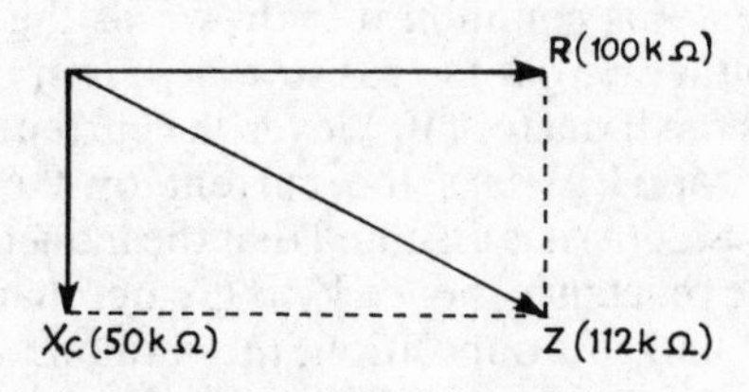

Fig. 8.23.

(b) The voltage across R (V_R) $= I \times R$

$$= 100 \times 10^{-6} \times 10^5$$

$$= 10\text{V}$$

The voltage across C (V_C) $= I \times X_C$

$$= 100 \times 10^{-6} \times 5 \times 10^4$$

$$= 5\text{V}$$

Phasors for V_R and V_C may now be drawn to scale as in Fig. 8.24 and the resultant measured to find V (11·2V).

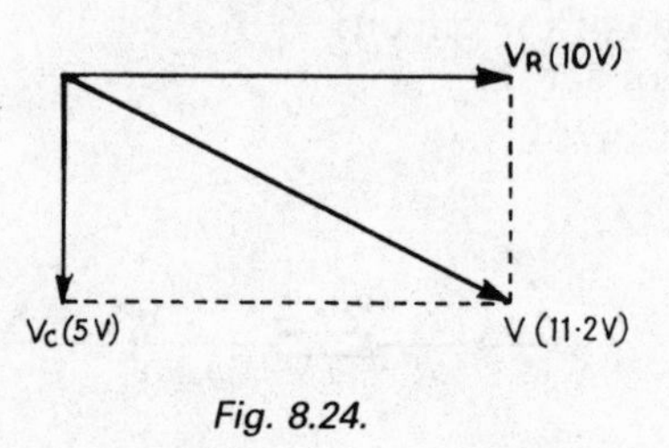

Fig. 8.24.

Circuits with *L*, *C* and *R* in series

Let us now consider a circuit where all three components are connected in series as in Fig. 8.25.

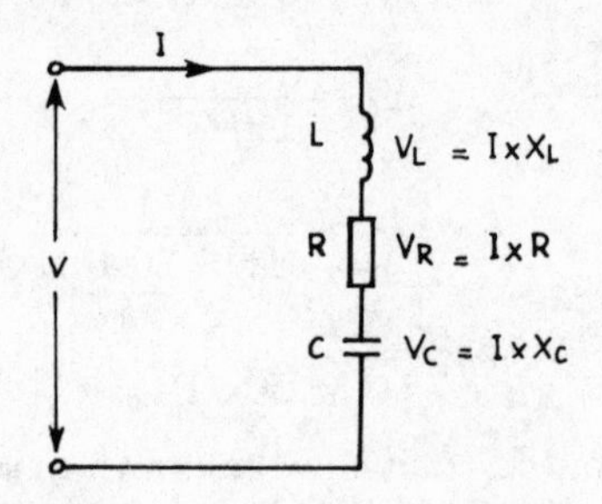

Fig. 8.25 L, C and R in series.

Since the current I is common it is drawn as the reference phasor, see Fig. 8.26. The voltage across the resistor (V_R) is in phase with the current, the voltage across the inductor (V_L) leads the current by 90° and the voltage across the capacitor (V_C) lags the current by 90°. Two conditions are illustrated; in Fig. 8.26(a) it is assumed that the inductive reactance is greater than the capacitive reactance, hence V_L is greater than V_C. Since the phasors for V_L and V_C act in direct opposition, the resultant of these two phasors is given by $(V_L - V_C)$. If the phasor addition of $(V_L - V_C)$ and V_R is now made, the resultant gives the applied voltage V.

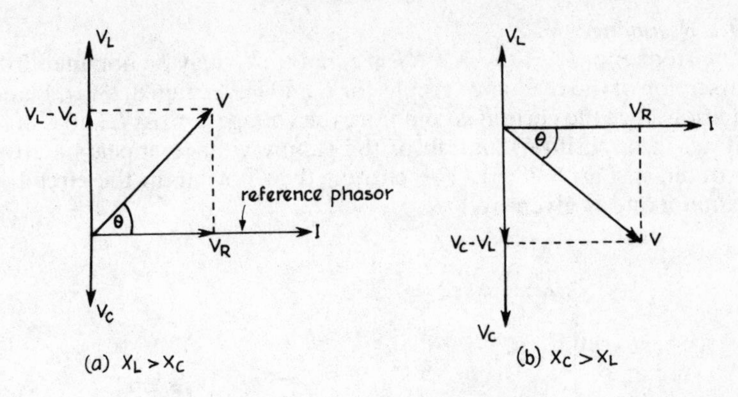

Fig. 8.26 Phasors for L, C and R in series.

Figure 8.26(b) shows the result when X_C is greater than X_L (which is the larger of the two depends upon the frequency). In this case the resultant of V_L and $_C$ is given by $(V_C - V_L)$. If the phasor addition of $(V_C - V_L)$ and V_R is now made, the resultant gives the applied voltage V. If the voltages across all three components are known, phasors may be constructed to scale and the resultant V measured. Alternatively the applied voltage may be found from:

$$V = \sqrt{[(V_L - V_C)^2 + V_R^2]}$$

Figure 8.27 shows how the impedance of the circuit may be obtained using

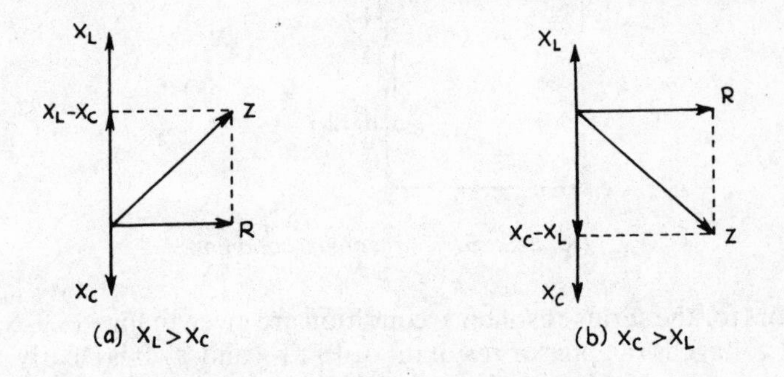

Fig. 8.27 Impedance of L, C and R in series.

phasors constructed to scale. Alternatively the impedance may be determined from:

$$Z = \sqrt{[(X_L - X_C)^2 + R^2]}$$

Assuming that the circuit values are constant there will be a particular frequency at which the reactance of L is exactly equal to the reactance of C. This results in an important condition knows as **series resonance**.

Series Resonance

The frequency (f_r) at which X_L is equal to X_C may be obtained from the intersection of the reactance graphs for L and C, see Fig. 8.28(a). Under this condition since the current is common, the voltages across L and C are equal (but act in opposition) and all of the supply voltage appears across R as illustrated in Fig. 8.28(b). The current then flowing in the circuit is at a **maximum** and is given by V/R.

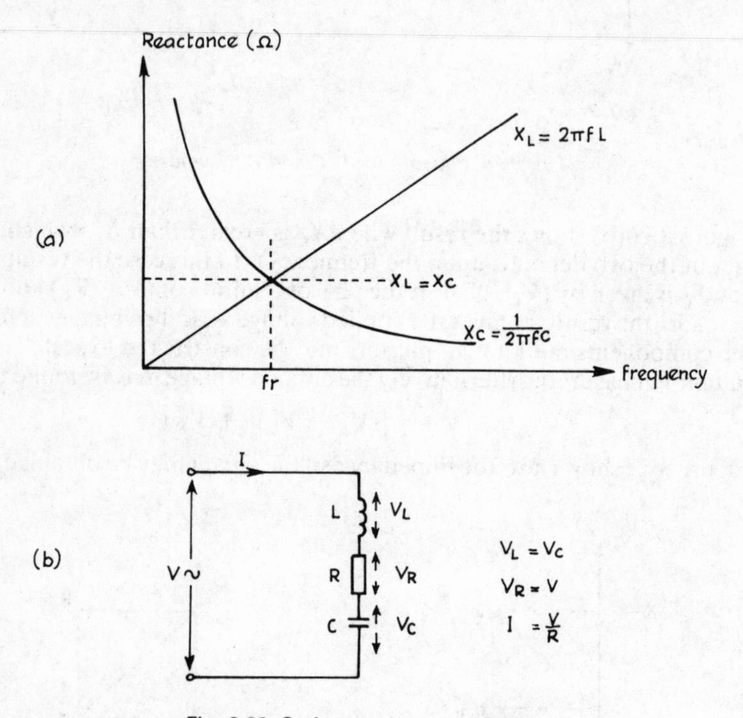

Fig. 8.28 Series resonant condition.

Phasors for the series-resonance condition are given in Fig. 8.29. Since the applied voltage is the phasor resultant of V_R, V_C and V_L it is clearly equal to V_R as the resultant of V_C and V_L is zero. **At resonance** therefore, the **current *I* is in phase with the applied voltage**, Fig. 8.29(a).

Impedance phasors are given in Fig. 8.29(b). Again, since Z is the phasor resultant of R, X_C and X_L, the impedance at resonance is purely resistive and equal to R because the resultant of X_L and X_C is zero. Since at the resonance frequency (f_r) $X_L = X_C$, the actual frequency at which resonance occurs may be solved from:

$$2\pi f L = \frac{1}{2\pi f C}$$

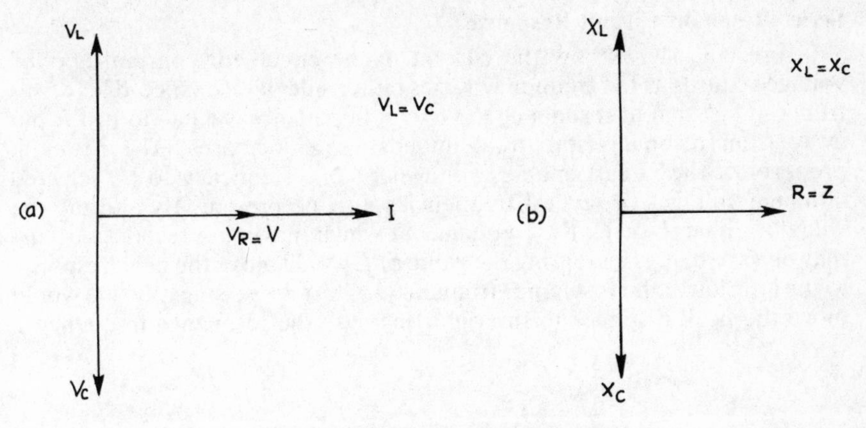

Fig. 8.29 Series resonance phasors.

$$\text{or } f^2 = \frac{1}{4\pi^2 LC}$$

$$f = \frac{1}{2\pi\sqrt{(LC)}} \text{ Hz}$$

Example 10

Calculate the series resonance frequency when:

(a) $L = 100\mu\text{H}$ and $C = 0.005\mu\text{F}$;
(b) $L = 5\mu\text{H}$ and $C = 100\text{pF}$

Solution

(a)

$$f_r = \frac{1}{2\pi\sqrt{(LC)}}$$

$$= \frac{1}{2\pi\sqrt{(100 \times 10^{-6} \times 0{\cdot}005 \times 10^{-6})}}$$

$$= \frac{10^6}{2\pi\sqrt{0{\cdot}5}}$$

$$= 225{\cdot}08\text{kHz}$$

(b)

$$f_r = \frac{1}{2\pi\sqrt{(LC)}}$$

$$= \frac{1}{2\pi\sqrt{(5 \times 10^{-6} \times 100 \times 10^{-12})}}$$

$$= \frac{10^9}{2\pi\sqrt{500}}$$

$$= 7{\cdot}12\text{MHz}$$

Series Resonant Circuit Response

Figure 8.30 shows how the current in the circuit for constant applied voltage reduces as the frequency varies either side of resonance. This arises from the fact that at resonance the circuit impedance is equal to just R but away from resonance the total impedance Z increases. The resonant property of the circuit enables voltages of one frequency to be selected although voltages of several frequencies may be present. By altering the value of either L or C, the frequency at which maximum response occurs may be varied, e.g. increasing the value of L would move the peak response to the left (lower the resonance frequency) whilst decreasing its value would move the peak response to the right (increase the resonance frequency).

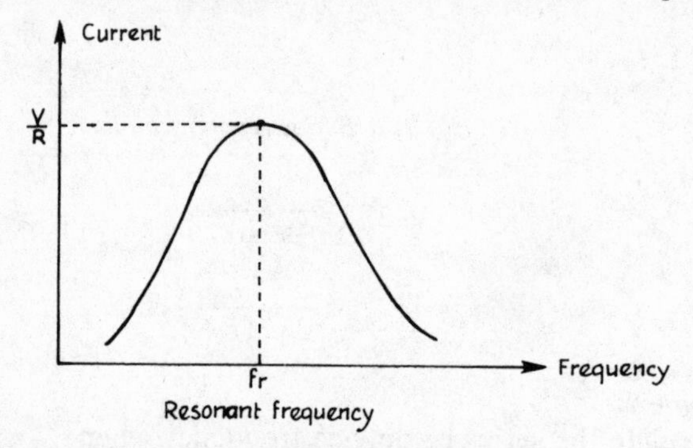

Fig. 8.30 Response of series resonant circuit.

A further important and interesting fact that occurs at resonance may be shown by simple calculation, see Fig. 8.31. Suppose that at resonance X_L and X_C are 100Ω, R is 5Ω and 20V is applied to the circuit.

Since at resonance $Z = R$, the current I at resonance is given by:

$$I = \frac{V}{R} = \frac{20}{5} = 4\text{A}$$

Voltage across $L = IX_L = 4 \times 100 = 400\text{V}$

Voltage across $C = IX_C = 4 \times 100 = 400\text{V}$

Voltage across $R = IR = 4 \times 5 = 20\text{V}$ = applied voltage

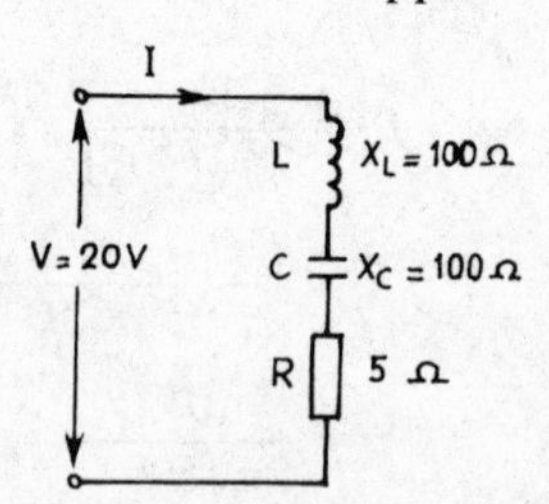

Fig. 8.31 Example of series resonance.

We now have what appears to be a very unusual situation in that the voltages across L and C are twenty times the applied voltage, referred to as the **magnification factor** (Q) of the circuit. This result is not in conflict with the rule that the voltages around a circuit will add up to the applied voltage, since the voltages across L and C are equal and opposite and their resultant is zero. A magnification factor of 100 or more may be met with in radio circuits.

If the resistance R of the circuit is increased, the current at resonance will be decreased, but the resistance will not have much effect at frequencies removed from resonance as the impedance is high. The effect of resistance on the response curve is shown in Fig. 8.32.

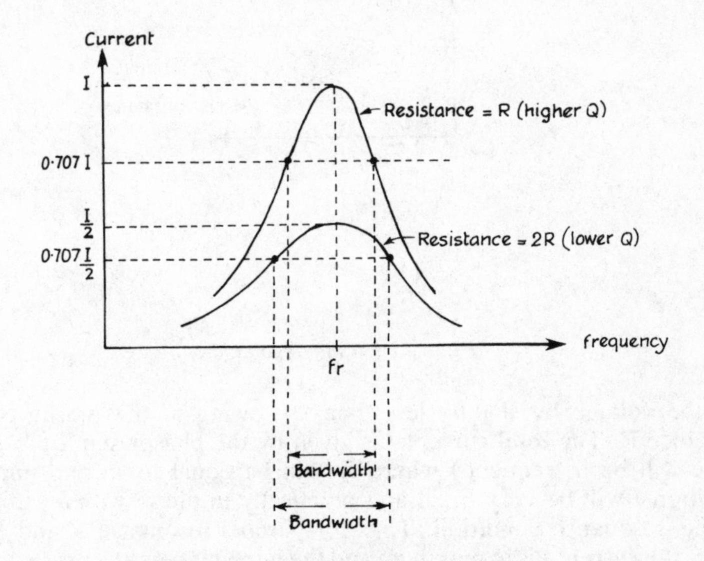

Fig. 8.32 Effect of resistance on resonance curve of series circuit.

Since the bandwidth of the resonant circuit is taken to be the frequencies covered between the points where the response has fallen to 0·707 of its maximum response (3dB points), the bandwidth of the circuit with the larger series resistance will be greater but the response will be less selective, i.e. of lower Q.

Parallel Resonant Circuit

Resonance may also be achieved when an inductor is connected in parallel with a capacitor as in Fig. 8.33(a). Since it is impossible to make an inductor without resistance, a resistor R is shown in series with L but does not normally exist as a physical component.

As the applied voltage V is common to both arms of the parallel circuit, it is drawn as the reference phasor, see Fig. 8.33(b). The current I_C in the capacitive arm leads the voltage by 90°. The current I_L in the inductive arm

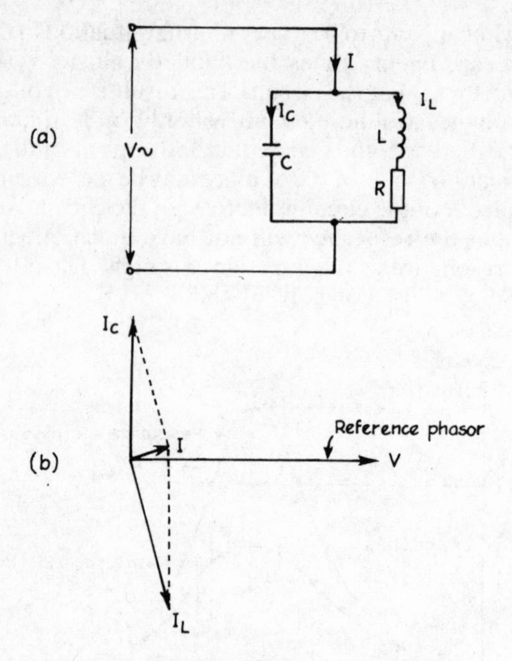

Fig. 8.33 Parallel resonance.

lags the voltage by slightly less than 90° owing to the presence of the resistance R. The total current I is given by the phasor sum of I_C and I_L. There will be a frequency where I_L will be equal to I_C and under this condition I will be very small and practically in phase with V; this is the parallel-resonance condition. Thus for parallel resonance V and I are in phase, the current I is at minimum and the impedance is at a maximum. The response curve of Fig. 8.34 shows how the current I and the impedance varies as the frequency varies either side of the resonance frequency.

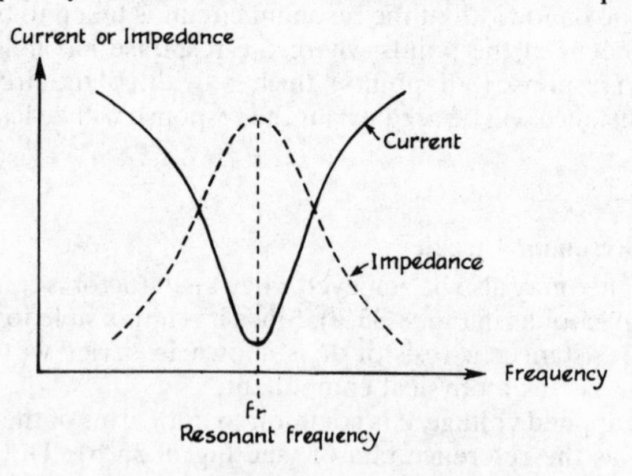

Fig. 8.34 Parallel resonance response.

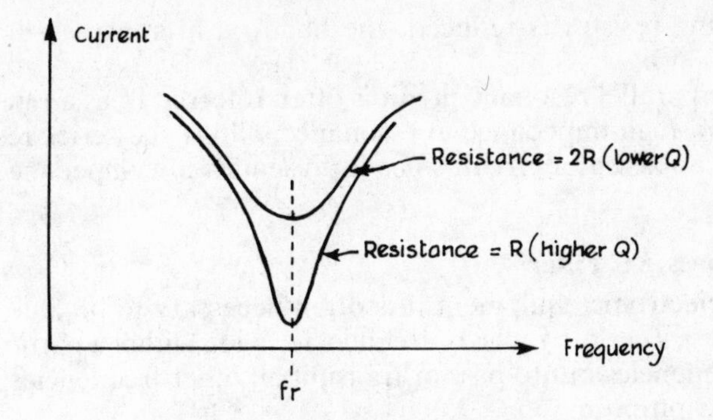

Fig. 8.35 Effect of resistance on resonance curve of parallel circuit.

The effect of increasing the resistance of the coil is shown in Fig. 8.35. As the resistance is increased the response becomes less selective but the circuit bandwidth is increased. In practice R is kept as small as possible. When R is small compared with the reactance of L, it may be shown that the resonance frequency is given by:

$$f = \frac{1}{2\pi\sqrt{(LC)}}\,\text{Hz}$$

the same as for the series-resonant circuit condition.

If the selectivity is too great, i.e. the bandwidth is too small for a particular application, the bandwidth may be increased by connecting a 'damping resistor' across the tuned circuit as shown in Fig. 8.36(a). As the value of the

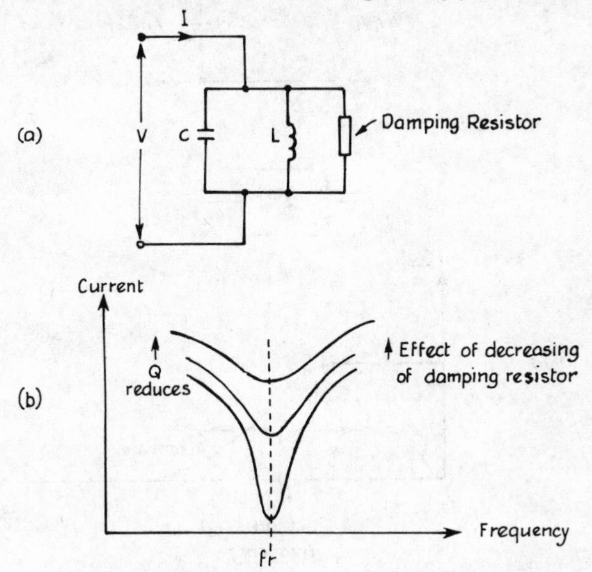

Fig. 8.36 Use of damping resistor to widen the bandwidth.

damping resistor is reduced, the bandwidth is increased as illustrated in Fig. 8.36(b).

The parallel resonant circuit is often referred to as a **rejector circuit** as it offers a high impedance at resonance, whilst the series resonant circuit is called an **acceptor circuit** since it presents a low impedance at resonance.

Low-pass *RC* Filter

In electronic equipment it is often necessary to be able to attenuate or suppress a certain range of frequencies lying within a particular larger band of frequencies but to pass or transmit all other frequencies; filters are used for this purpose.

A low-pass *RC* filter is shown in Fig. 8.37(a) and its symbol in Fig. 8.37(b). A low-pass filter passes low-frequency signals (ideally without loss) but severely attenuates signals of above a certain frequency. Fig. 8.37(c) shows the response where the ratio v_o/v_i is plotted against frequency. Over the range where $v_o/v_i = 1$, the output v_o is equal to the input voltage v_i and here the filter is passing signal frequencies without loss. As the *RC* filter response changes slowly from pass to attenuate, it is necessary to specify a frequency where for practical purposes the filter changes over from pass to attenuate. This frequency, called the **cut-off frequency** f_{co}, is taken to be the frequency where the output voltage has fallen to 0·707 (3dB) of the input voltage. Thus

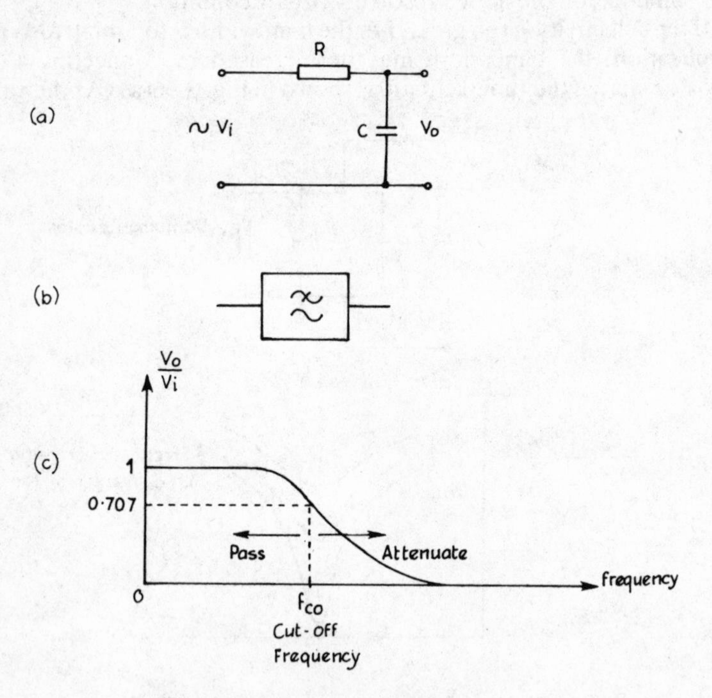

Fig. 8.37 Low-pass R–C filter.

above f_{co} the low-pass filter is attenuating the input signal as the ratio v_o/v_i is less than 0·707.

The action of the low-pass RC filter may be considered as that of a potential divider, see Fig. 8.38. If the value of C is chosen so that it has a very high reactance compared with the resistance R at low frequencies, then most of the input voltage will appear across C (V_C) and little voltage will appear across R (V_R). Since V_C is the output voltage V_o, then at low frequencies the filter will be operating in the pass-band. As the frequency is raised the reactance of C falls resulting in less voltage across C but more across R. When the reactance of C is very small compared to R, there will be an extremely small voltage across C with practically all of the input voltage developed across R; here the filter is in the attenuation band (above f_{co}).

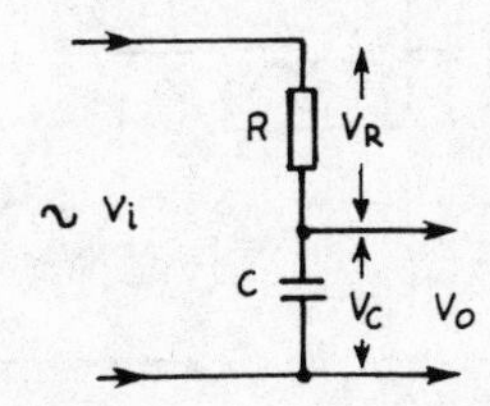

Fig. 8.38 Potential divider action of L.P.F.

The cut-off frequency of the filter depends upon the time constant of the circuit which is determined by the CR product. At f_{co} the voltages across R and C will be equal but V_C will lag V_R by 90°.

$$\text{Thus, at } f_{co},\ R = \frac{1}{2\pi f C}\ \Omega$$

$$\text{or } f_{co} = \frac{1}{2\pi CR}\ \text{Hz}$$

Example 11

In a low-pass RC filter, $R = 50\text{k}\Omega$ and $C = 100\text{nF}$. Determine the cut-off frequency.

Solution

$$f_{co} = \frac{1}{2\pi CR}\ \text{Hz}$$

$$= \frac{1}{2\pi \times 100 \times 10^{-9} \times 5 \times 10^4}$$

$$= 31{\cdot}8\text{Hz}$$

Thus the filter will pass frequencies from 0Hz to 31·8Hz but attenuate frequencies above 31·8Hz.

High-pass *RC* Filter

If the positions of R and C of the low-pass filter are interchanged, a high-pass filter results, see Fig. 8.39(a). The circuit symbol is given in Fig. 8.39(b) and the filter response in Fig. 8.39(c).

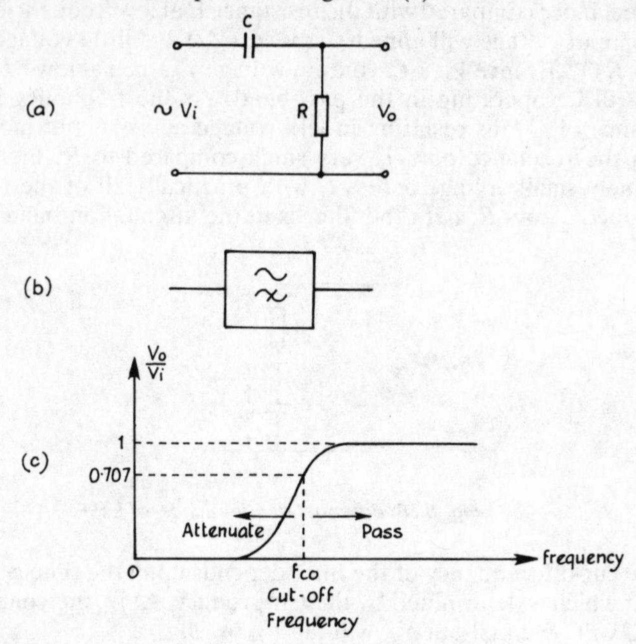

Fig. 8.39 High-pass R–C filter.

The filter passes frequencies down to the cut-off frequency f_{co} and below this frequency the filter changes over from pass to attenuate. It will be seen that the high-pass filter has a response which is the mirror image of the low-pass filter response.

Again the action of the filter may be considered as a potential divider, see Fig. 8.40. At high frequencies where the reactance of C is quite small compared with the resistance R, V_R is large compared with V_C and the filter is in the pass-band. If the frequency is reduced, the reactance of C rises resulting in more voltage across C and less across R. Below f_{co} where the

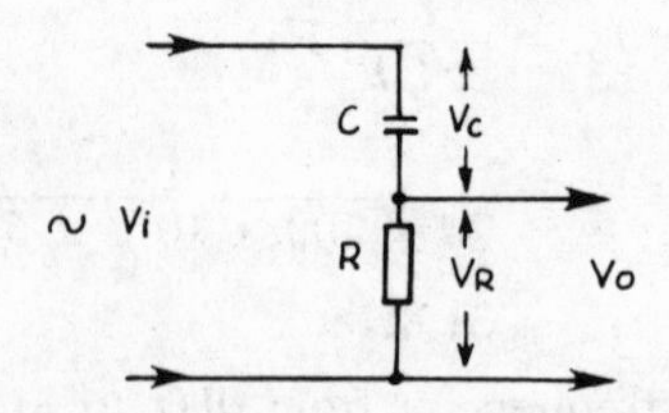

Fig. 8.40 Potential divider action of H.P.F.

reactance of C is very large compared with R, most of the input voltage appears across C with very little across R; here the filter is operating in the attenuation band.

The cut-off frequency of the filter again depends upon the CR product and f_{co} is given by:

$$f_{co} = \frac{1}{2\pi CR} \text{ Hz}$$

Thus if the same values as given in Example 11 are used for the high-pass filter, the filter will pass frequencies above 31·8Hz but attenuate frequencies below this figure.

QUESTIONS ON CHAPTER EIGHT

(1) A capacitor of 2000pF is charged via a resistance of 5kΩ to a d.c. supply of 20V. The time taken for the voltage across the capacitor to reach 14·14V will be:
(a) 1·414μs
(b) 0·707μs
(c) 1·414ms
(d) 4μs.

(2) A capacitor is charged via a resistance to a d.c. supply of 15V. After a period equal to the time constant the voltage across the capacitor will be:
(a) 5·0V
(b) 5·55V
(c) 9·45V
(d) 15V.

(3) A capacitor of 5μF charged initially to 100V is discharged via a resistance of 100kΩ. The approximate time for the capacitor voltage to reach 0V will be:
(a) 2·5s
(b) 0·5s
(c) 100μs
(d) 50ms.

(4) An inductor of 5mH is connected in series with a resistance of 50kΩ across a 50V d.c. supply. The time taken for the current in the circuit to grow to 0·63mA will be:
(a) 0·1μs
(b) 250s
(c) 10μs
(d) 2·5s.

(5) The reactance of an inductor is given by:
(a) $2fL$
(b) $\frac{1}{2\pi fL}$
(c) $2\pi L$
(d) $2\pi fL$.

(6) The reactance of a capacitor is given by:
(a) $2\pi fC$
(b) $\frac{1}{\omega C}$
(c) $\frac{2\pi f}{C}$
(d) πfC.

(7) When a sinewave voltage is applied to a circuit consisting of an inductor in series with a resistor:
(a) The current in the inductor leads the current in the resistor by 90°
(b) The voltage across the resistor leads the current by 90°
(c) The voltage across the inductor leads the voltage across the resistor by 90°
(d) The voltage across the inductor and resistor are in-phase.

(8) When a sinewave voltage is applied to a circuit consisting of a capacitor in series with a resistor:
(a) The voltage across the capacitor leads the current by 90°
(b) The voltage across the resistor and capacitor are in-phase
(c) The applied voltage and the current are in-phase
(d) The voltage across the resistor leads the voltage across the capacitor by 90°.

(9) In a series resonant circuit:
(a) The impedance at resonance is high
(b) The applied voltage and current are in-phase at resonance
(c) The voltage across the inductor and capacitor are 90° out of phase
(d) Minimum current flows at resonance.

(10) In a parallel resonant circuit:
(a) Maximum current flows at resonance
(b) The impedance at resonance is high
(c) The impedance at resonance is low
(d) The applied voltage and current are 90° out of phase.

(11) The resonance frequency of a series tuned circuit is given by:
(a) $f = \frac{\sqrt{(LC)}}{2\pi}$

(b) $f = \frac{2\pi\sqrt{L}}{C}$

(c) $f = \frac{1}{2\pi LC}$

(d) $f = \frac{1}{2\pi\sqrt{(LC)}}$.

(12) The effect of connecting a resistor across a parallel tuned circuit is to:

(a) Alter the resonance frequency
(b) Increase the bandwidth
(c) Decrease the bandwidth
(d) Decrease the current at resonance.

(Answers on page 180)

CHAPTER NINE

TRANSFORMERS AND SHIELDING

TRANSFORMERS

WHEN AN ALTERNATING current flows in an inductor there is a magnetic field around the inductor of strength that is proportional to the current flowing, see instants a, b, c and d of Fig. 9.1. Thus with a.c. there is a magnetic field that is either growing or contracting and therefore **moving**. It is the moving magnetic field which cuts the turns of the inductor and induces an e.m.f. (self-induced e.m.f.) into those turns. It therefore seems reasonable that if a second coil is brought close to the first so that the moving magnetic field cuts the turns of the second coil an e.m.f. will be induced into the second coil; this is the principle of the transformer.

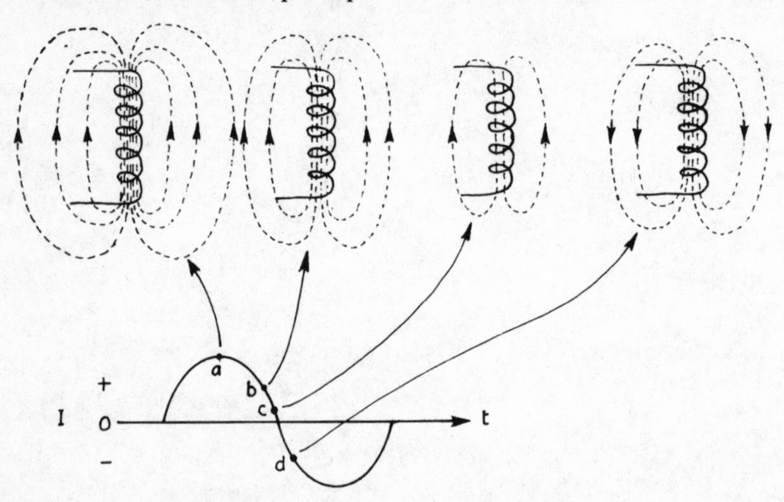

Fig. 9.1 Magnetic field associated with an inductor carrying alternating current.

In most cases the two coils, called the primary and secondary are wound on a magnetic core (iron), as illustrated in Fig. 9.2. This has two effects: (i) it

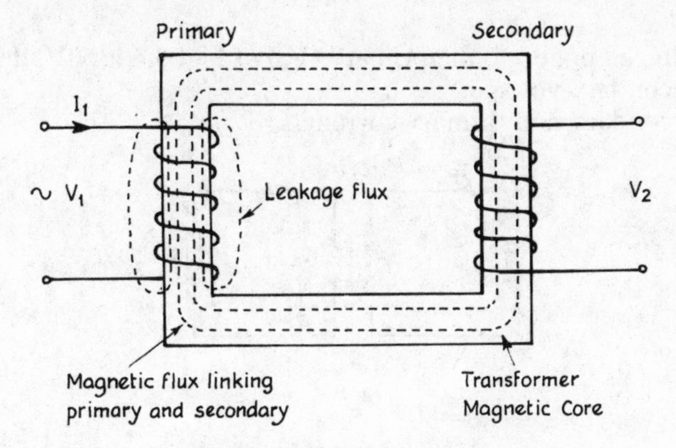

Fig. 9.2 Transformer principle.

increases the inductance of the primary and so reduces the current that flows in the primary and (ii) it concentrates the magnetic flux so that most of it passes through the secondary winding (the portion of flux not linking with the secondary is called the leakage flux). Thus, when an a.c. supply of voltage $V1$ is applied to the primary, a current I_1 will flow. This current sets up a moving magnetic field which cuts the turns of the secondary inducing a voltage $V2$ in it. The secondary induced e.m.f. will follow the same changing pattern of the primary applied voltage.

Voltage and Current Ratios

Let us consider an 'ideal' transformer, i.e. one with no losses and one where all of the flux set up in the primary links with the secondary. If a load R is connected across the secondary then the secondary induced e.m.f. V_s will produce a current in the secondary I_s of value equal to V_s/R, see Fig. 9.3. If the ratio of the number of primary turns to the number of secondary turns is N_p/N_s then:

$$\frac{V_p}{V_s} = \frac{N_p}{N_s} \text{ and } \frac{I_p}{I_s} = \frac{N_s}{N_p}$$

i.e. the ratio of the voltages across primary and secondary is proportional to the turns ratio, whereas the ratio of currents in primary and secondary is inversely proportional to the turns ratio.

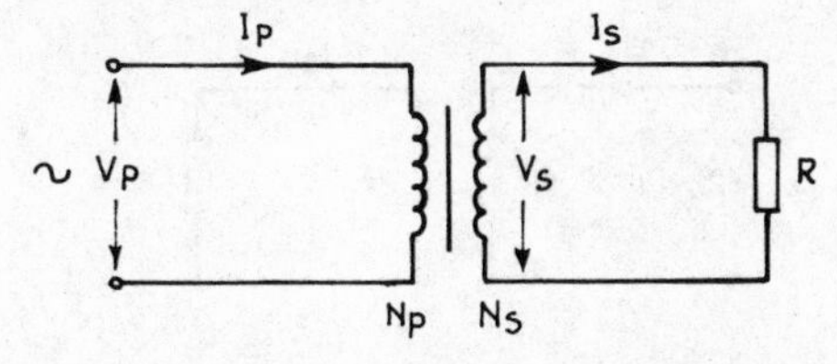

Fig. 9.3 Voltage and current ratios (ideal transformer).

Example 1

If the voltage applied to the primary (V_p) of Fig. 9.4 is 50V, determine:
(a) the secondary voltage;
(b) the secondary and primary currents.

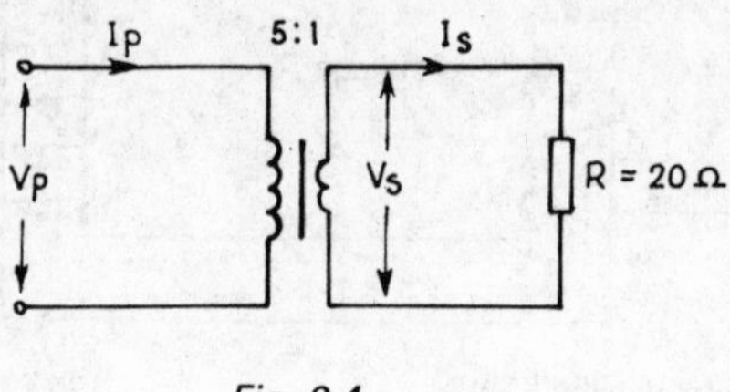

Fig. 9.4.

Solution

(a)
$$\text{Now} \quad \frac{V_p}{V_s} = \frac{N_p}{N_s} = \frac{5}{1}$$

$$\therefore \quad \frac{50}{V_s} = \frac{5}{1}$$

$$\text{or} \quad V_s = \frac{50}{5} = 10\text{V}$$

(b)
$$I_s = \frac{V_s}{R} = \frac{10}{20} = 0.5\text{A}$$

$$\text{Now} \quad \frac{I_p}{I_s} = \frac{N_s}{N_p} = \frac{1}{5}$$

$$\therefore \quad \frac{I_p}{0{\cdot}5} = \frac{1}{5}$$

$$\text{or} \quad I_p = \frac{0.5}{5} = 0.1\text{A}$$

Impedance Ratio

Consider Fig. 9.5 where R_p is the resistance or impedance 'seen' looking into the primary and let $N_p/N_s = n{:}1$.

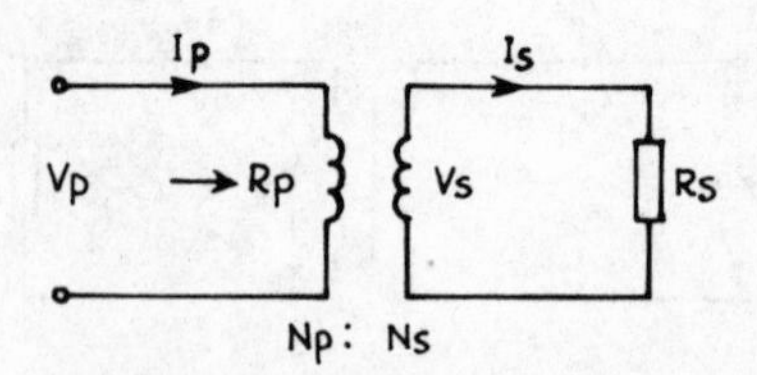

Fig. 9.5 Impedance ratio.

$$\text{Now } R_p = \frac{V_p}{I_p} \text{ and } R_s = \frac{V_s}{I_s}$$

$$\text{thus } \frac{R_p}{R_s} = \frac{V_p}{I_p} \div \frac{V_s}{I_s}$$

$$= \frac{V_p}{I_p} \times \frac{I_s}{V_s} = \frac{V_p}{V_s} \times \frac{I_s}{I_p}$$

$$\text{but } \frac{V_p}{V_s} = \frac{N_p}{N_s} = \frac{I_s}{I_p} = n$$

$$\therefore \frac{R_p}{R_s} = n \times n = n^2$$

$$\text{or } R_p = n^2 R_s$$

Thus the impedance looking into the primary is equal to the (turns ratio)^2 × the secondary impedance.

Example 2

Determine the primary impedance of Fig. 9.6 (a) and (b).

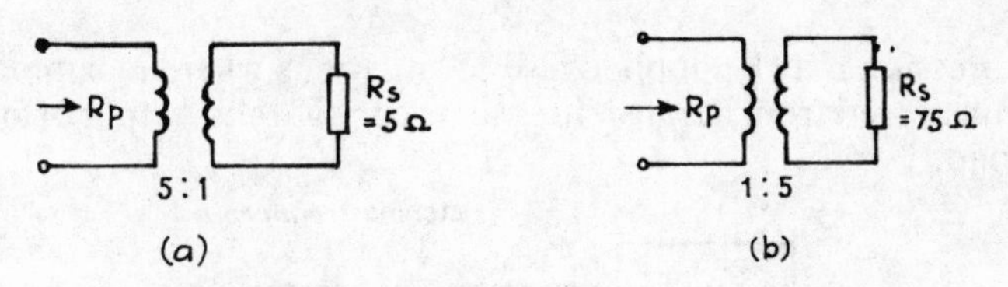

Fig. 9.6.

Solution

(a)

$$R_p = n^2 R_s \ (n{:}1 = \text{step-down ratio})$$
$$= 25 \times 5 = 125\Omega$$

(b) Since a step-up ratio is used here $(1{:}n)$,

$$\frac{R_p}{R_s} = \frac{1}{n^2}$$

$$\text{or } R_p = \frac{R_s}{n^2} = \frac{75}{25} = 3\Omega$$

Transformer Matching

If we have a signal source such as an amplifier and a load, e.g. a loudspeaker and we wish to transfer maximum power from the amplifier to the loudspeaker, then the loudspeaker resistance or impedance should be

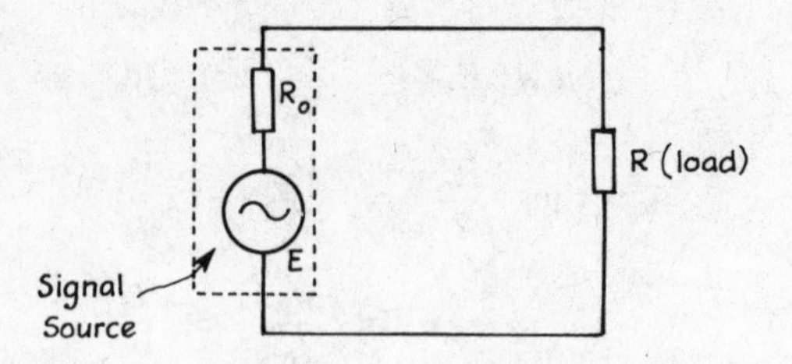

Fig. 9.7 Maximum power transfer.

equal to the output resistance or output impedance of the amplifier. This principle is illustrated in Fig. 9.7 where, for maximum power transfer from the signal source, R should be equal to R_o.

Usually the resistances are not equal and in such cases a transformer may be used for impedance matching, as shown in Fig. 9.8. When R and R_o are known the turns ratio required may be determined from

$$n^2 = \frac{R_o}{R}$$

$$\text{or } n = \sqrt{\frac{R_o}{R}}$$

The expression may be applied also to situations where maximum power is to be transferred from a transmitter to its aerial or the output of an oscillator to its load.

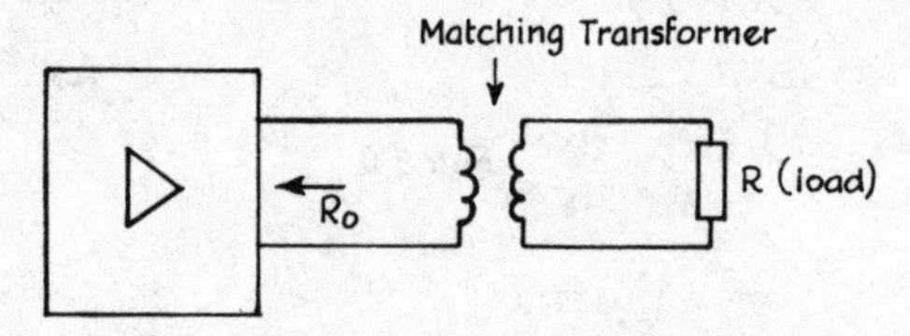

Fig. 9.8 Use of matching transformer.

Example 3

An amplifier has an output resistance of 4.5kΩ and is to be matched to a 20Ω loudspeaker for maximum power transfer. Determine the turns ratio of the transformer to be used.

Solution

$$n = \sqrt{\frac{R_o}{R}}$$

$$= \sqrt{\frac{4500}{20}}$$

$$= \sqrt{225}$$

$$= 15 : 1 \text{ (step-down)}$$

Auto-transformer

The transformer illustrated in Fig. 9.2 is sometimes called a 'double-wound' transformer, i.e. having two windings. A transformer may also be constructed using a single winding with one or more tappings. The diagram of Fig. 9.9 shows an auto-transformer with a single tapping. The voltage

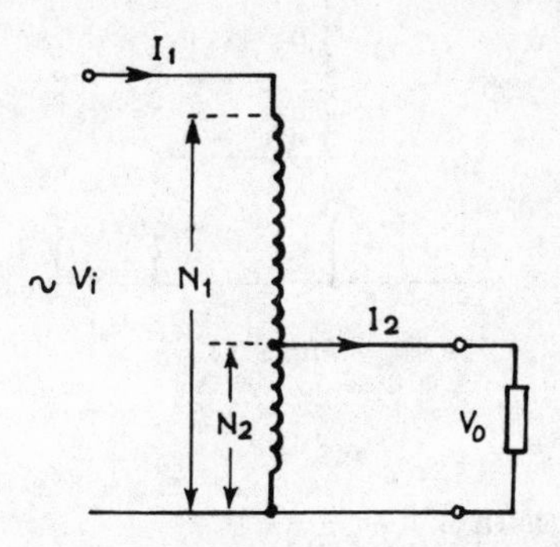

Fig. 9.9 Auto-transformer with single tapping.

ratio is determined in the same way as for the double-wound transformer, i.e.

$$\frac{V_i}{V_o} = \frac{N1}{N2} \text{ and the current ratio by}$$

$$\frac{I_1}{I_2} = \frac{N2}{N1}$$

An auto-transformer with several tappings is shown in Fig. 9.10. The voltage ratios are given by:

$$\frac{V_i}{V_{o_1}} = \frac{N1}{N4}, \quad \frac{V_i}{V_{o_2}} = \frac{N1}{N3} \text{ and } \frac{V_i}{V_{o_3}} = \frac{N1}{N2}$$

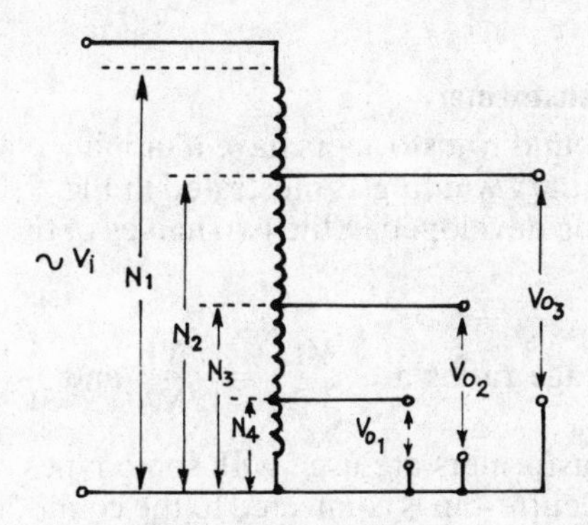

Fig. 9.10 Auto-transformer with more than one tapping.

Example 4

Sections a, b and c of the auto-transformer shown in Fig. 9.11 each have 500 turns. Determine the output voltages $V1$ and $V2$.

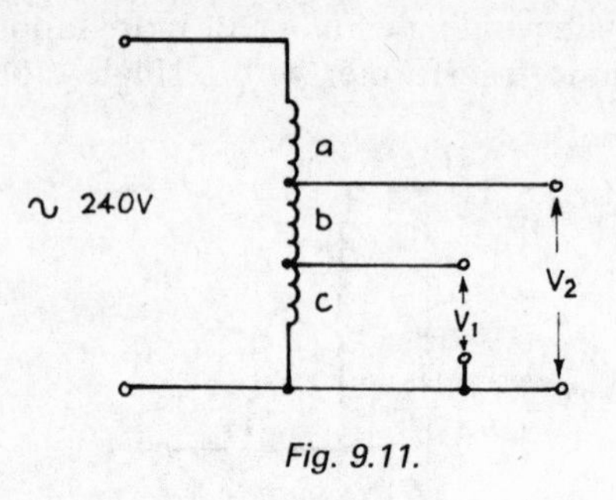

Fig. 9.11.

Solution

$$\text{Turns ratio for } V1 = \frac{1500}{500} = \frac{3}{1}$$

$$\therefore \frac{240}{V1} = \frac{3}{1}$$

$$\text{or } 3V1 = 240$$

$$V1 = 80\text{V}$$

$$\text{Turns ratio for } V2 = \frac{1500}{1000} = \frac{3}{2}$$

$$\therefore \frac{240}{V2} = \frac{3}{2}$$

$$\text{or } 3V2 = 480$$

$$V2 = 160\text{V}$$

Centre-tapped Transformer

Some double-wound transformers have a tapping placed at the electrical centre of the secondary winding as illustrated in Fig. 9.12(a). This results in equal voltages being developed in the two halves of the secondary winding, i.e. $V2 = V3$.

$$\text{The voltage ratios are } \frac{V1}{V2} = \frac{N1}{N2} \text{ and } \frac{V1}{V3} = \frac{N1}{N3}$$

Centre-tapped transformers are used with some types of full-wave rectifier where usually the centre-tap is connected to the common line or earth. Such a transformer may be designated as 300V–0–300V which means that there is

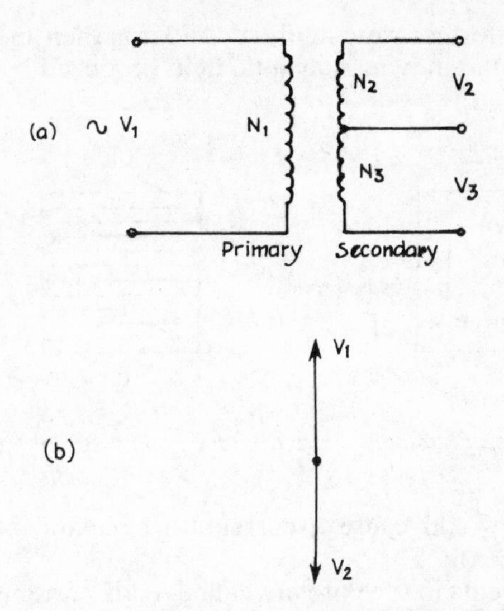

Fig. 9.12 Centre-tapped transformer.

300V (r.m.s.) across each of the two halves of the secondary. This type of transformer is also used in phase discriminators due to its ability to supply equal balanced voltages.

If the secondary winding is continuous then the voltages across each half will be in-phase with one another. However, **with respect to the centre-tap** the voltages are in anti-phase with one another and may be represented by the phasor diagram of Fig. 9.12(b).

Transformer Losses

An 'ideal transformer' has no losses and the power developed in the secondary is equal to the power supplied to the primary, or $V_p \times I_p = V_s \times I_s$. Practical transformers exhibit various power losses which means that less power is developed in the secondary than is supplied to the primary. The principal losses are:

(a) Copper Losses

The windings in a transformer are made from copper and as such each winding will have a d.c. resistance. Thus when current flows in a winding, there will be some power dissipated in the d.c. resistance of the winding. The effect of the primary winding d.c. resistance means that there will be effectively less voltage developed in the secondary and the effect of the secondary winding d.c. resistance will result in less voltage being supplied to the secondary load. Thus there will be a power loss between primary and secondary.

(b) Core Losses

If the core of a transformer were made of solid iron then an e.m.f. would be induced into it by the moving magnetic field produced by the primary

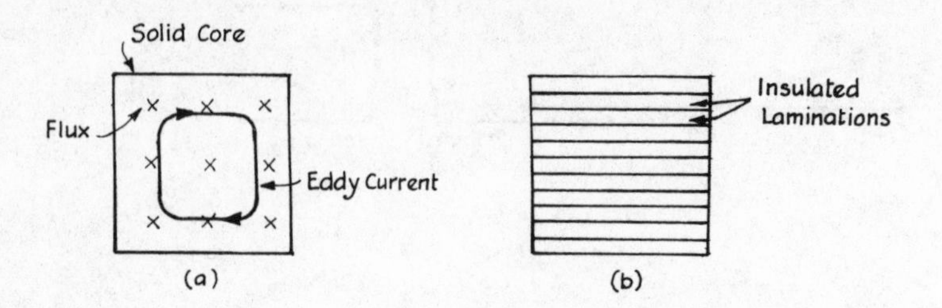

Fig. 9.13 Use of laminated core to reduce eddy-current loss.

winding. The e.m.f. would cause a current to circulate in the core as illustrated in Fig. 9.13(a).

The circulating currents in the core are called 'eddy currents' and would cause the core to get hot resulting in an undesirable power loss. To reduce the effect to small proportions, the core is made of thin iron laminations which are lightly insulated from one another as shown in Fig. 9.13(b). Currents will flow in the laminations, but these will be small as the flux passing down each lamination is small.

(c) Leakage flux

The useful flux in a transformer links both windings. Leakage flux links with its own turns but not both windings and its path exists largely in air. It is really wasted flux and represents a loss. Leakage flux is reduced by careful design and construction.

Practical transformers have efficiencies in the range 96–99 per cent.

Types of Transformer

Mains Transformers

These transformers operate from the mains supply of 240V r.m.s. at 50Hz and are used to supply appropriate voltages to operate integrated circuits and discrete transistors in various electronic equipment. Since modern electronic circuits require supply voltages in the range 6–50V, the transformer used is a step-down one.

The primary and secondary windings are not wound on separate limbs as drawn in theoretical diagrams. They are wound in intimate contact so that the flux induced by one winding links with the other, as illustrated in Fig. 9.14 where they are wound over the centre limb of the low-loss iron core. A core which is subjected to alternating flux is built up from laminations to reduce eddy current losses using often T and U laminations,

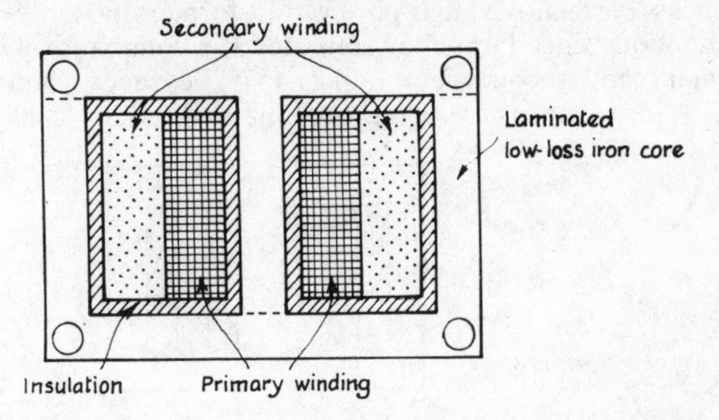

Fig. 9.14 Mains transformer construction.

see Fig. 9.15(a). Silicon iron is used in most modern power transformers, the addition of about four per cent of silicon to pure iron increasing its resistivity by about four times, so that the eddy current loss is reduced. Alternate groups of three to four laminations are overlapped as shown in Fig. 9.15(b) to produce a low reluctance joint thereby reducing the leakage flux that would be present with a butt joint. The silicon-iron core is normally connected to earth.

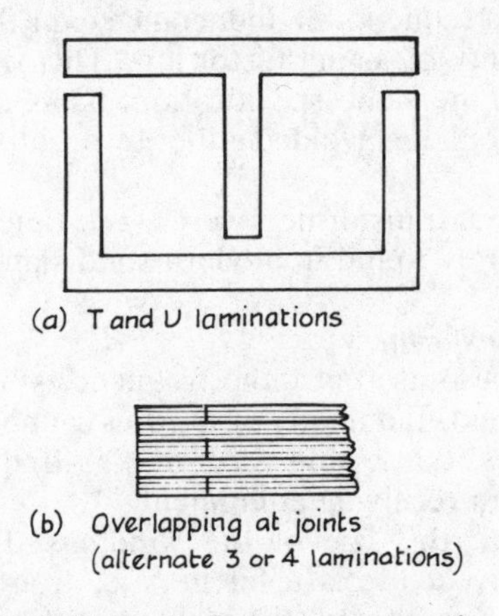

Fig. 9.15 Laminated core construction.

In a mains transformer where considerable power output is required, the insulated copper windings must be of sufficient diameter to carry the current without overheating. Thus the larger the power output, the larger the transformer.

Audio-frequency Transformers

Transformers of similar construction are also used at audio frequencies. Low-power audio transformers may use a core material of high permeability such as Permalloy. High-power transformers generally use silicon iron laminations since Permalloy saturates at a comparatively low flux density. Primary and secondary windings may be sandwiched as illustrated in Fig. 9.16 to improve winding coupling and reduce leakage flux.

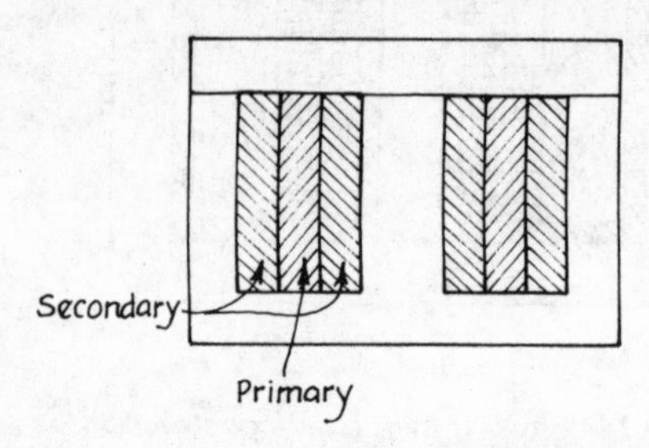

Fig. 9.16 Winding arrangement for a.f. transformer.

The design of this type of transformer is more difficult than a mains transformer due to the need to achieve an even response over a band of frequencies, e.g. 30Hz to 15kHz. The size of the primary inductance limits the response at low frequencies whilst the self-capacitance and leakage inductance of the windings are responsible for response limitations at high frequencies. In addition, if one of the windings carries a d.c. component of current it will cause a reduction in inductance over that attainable without d.c. and the possibility of signal distortion. This would require a larger transformer to meet the same specification. Thus unbalanced d.c. in an audio transformer should be avoided either by push-pull operation or shunt feeding.

Interstage and output transformers were at one time used with valve audio amplifiers but are rarely found in modern solid-state circuits.

Radio-frequency Transformers

Transformers are also used at radio frequencies but the construction is generally quite different. Laminated iron cores cannot be used owing to the large eddy current loss that would occur at high frequencies and the small signals encountered in receiving equipment.

In order to reduce the size of the windings, Ferroxcube cores are commonly used. Ferroxcube is a ferrite, i.e. iron oxide having a high permeability and high resistivity (the high resistivity reduces eddy current losses). In most cases, the primary and secondary are wound as shown in Fig. 9.17(a) on an insulating tube with a solid cylinder of Ferroxcube fitted inside. In receiving equipment the primary and secondary windings may form resonant circuits and when a Ferroxcube core is used the inductance and hence the resonant frequency may be varied by moving the core in and out of the winding as in Fig. 9.17(b). Movement of the Ferroxcube core into the winding will increase the inductance and hence reduce the resonant frequency. Often no core is used (air-cored) and the windings may be placed side-by-side or on top of one another as shown in Fig. 9.17(c).

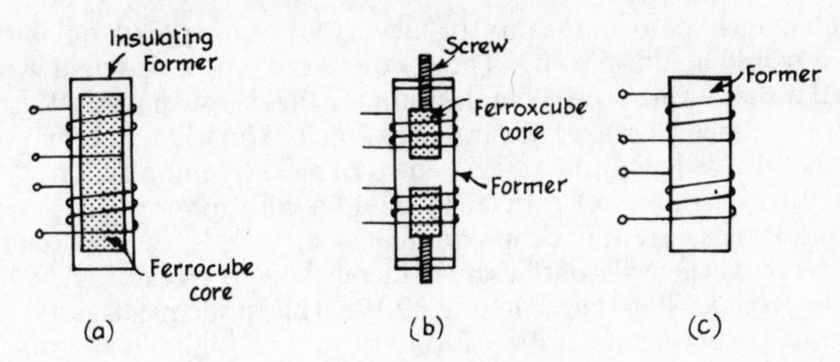

Fig. 9.17 R.F. transformers.

SHIELDING

Magnetic Shielding

Consider a coil carrying current and the resultant magnetic field around the coil spreading out into the space surrounding it as in Fig. 9.18(a). If the current in the coil is **d.c.** or **low-frequency a.c.** and we wish to prevent the flux lines from intercepting other neighbouring components, the coil may be enclosed in a screening box of high permeability material such as **Mumetal** as shown in Fig. 9.18(b). The walls of the screening box form an easy

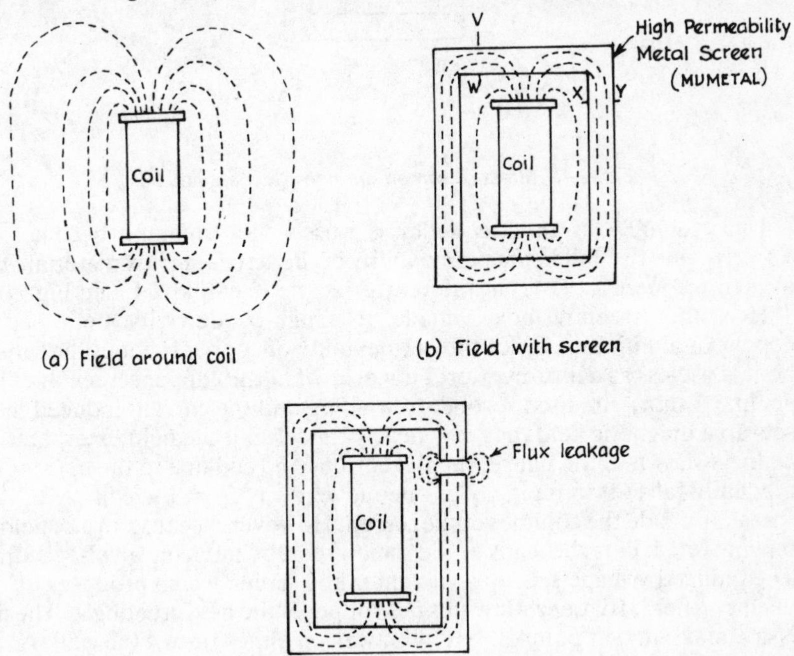

Fig. 9.18 Magnetic screening at d.c. and low frequencies.

(low-reluctance) path for the flux so that very little flux spreads out into areas where it would be unwelcome. The action of the screen or shield would be nullified if there were a break in the path for the flux such as V–W or X–Y. The result of such a break is illustrated in Fig. 9.18(c) where flux leaks at the break thereby reducing the effectiveness of the screening action.

It may be desirable, of course, to shield a coil from external magnetic fields and if these are d.c. or low-frequency a.c. fields, the low reluctance path offered by the walls of the screen flux will divert an external field away from the coil as illustrated in Fig. 9.19. This principle is used in the low-frequency screening of transformers, chokes, cathode ray tubes and measuring instruments.

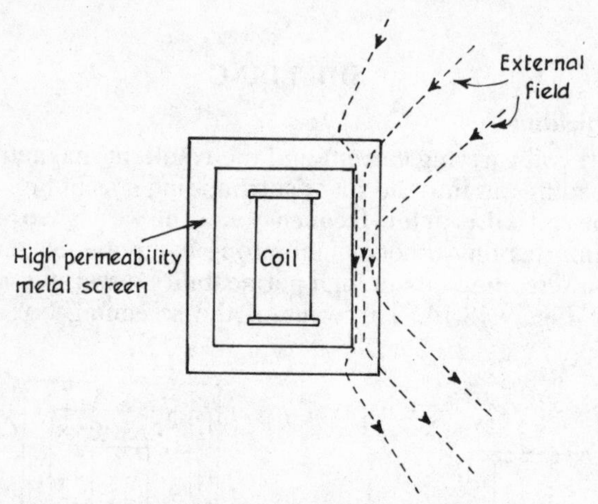

Fig. 9.19 Effect of screen on external magnetic field.

Unfortunately as the frequency is raised the screening becomes less effective due to a fall in the permeability of the screening box material. Thus at high frequencies (r.f.) a different principle is employed, see Fig. 9.20.

Here the screening box is made of a high-conductivity metal such as copper or aluminium. Since the permeability of copper is about the same as air it is useless as a flux diverter. However what the copper screen does is to act like a short-circuited secondary winding and the current induced into it sets up a magnetic field that very nearly cancels out the field external to the coil. As shown in the figure, the flux around the coil due to the current in it gradually falls off in intensity as one moves away from the coil (ABC) and spreads outside the confines of the screen. However, because an a.c. field is a moving one it cuts the walls of the screen thereby inducing a voltage into it. The induced voltage sets up a current in the screen which produces its own magnetic field. By Lenz's law this field opposes the field creating it. The field is at a maximum at point B′ but falls away on either side of the wall, A′ and C′, resulting external to the screen two fields acting in opposite directions which **very nearly cancel one another**.

It will be noted that there is cancellation of the magnetic field due to the

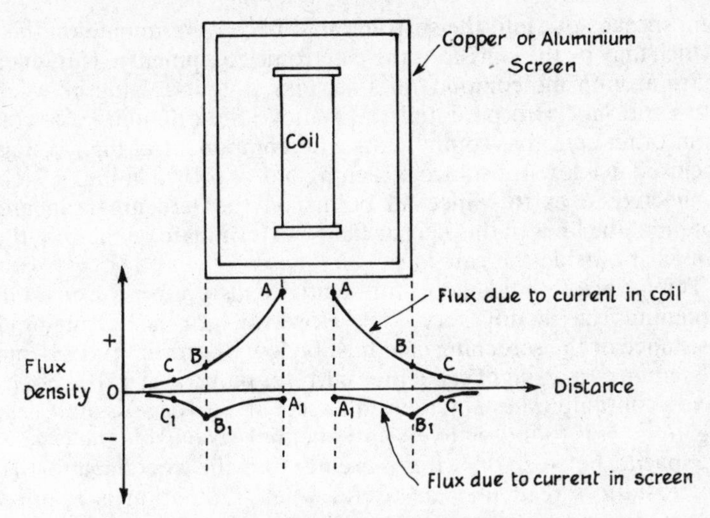

Fig. 9.20 Magnetic screening at high frequencies.

coil inside the screen, i.e. the effect of the screen is to reduce the inductance of the coil. To avoid a large reduction in inductance the space inside the screen should be as large as is practicable.

The principle holds good in shielding the coil from external r.f. magnetic fields.

Electrostatic Screening

No mention has been made of an 'earth' in connection with magnetic screening because earthing has nothing to do with magnetic screening but it has everything to do with electrostatic screening. Consider a component at high potential ($+V$) with respect to earth as in Fig. 9.21(a). The electrostatic

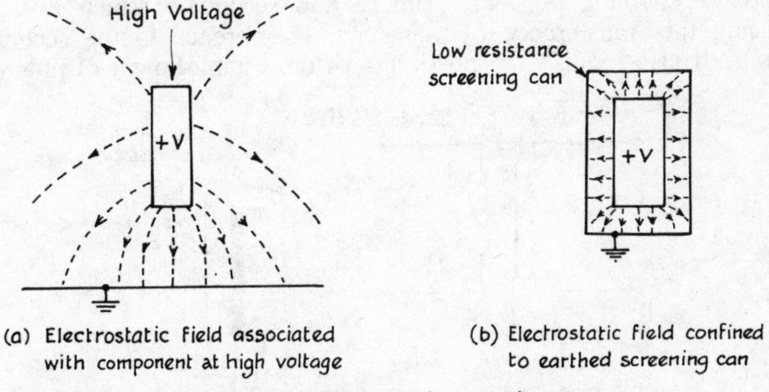

Fig. 9.21 Electrostatic screening.

field spreads out into the surrounding space to terminate on the earth line (which may be the chassis of the electronic equipment). Thus every point in space around the component is at some potential lying between $+V$ and earth, i.e. not zero potential. To remove the offending electrostatic field from other sensitive components, the component at high voltage may be enclosed in a low-resistance screening box as shown in Fig. 9.21(b) which is connected to earth. Since all points on the screening can are at earth potential the lines of the electric field will terminate on the can, thus no field will exist outside the can.

This principle holds good for steady fields even if the resistance of the screening can is not very low. However, for fast-changing fields the resistance of the screening can must be low to prevent p.d.s being set up in the screen as a result of capacitive currents induced in it. It is not essential to have a continuous metal screen and a sort of 'bird cage' is quite effective, see Fig. 9.22. It is found that very little field exists outside the cage even when the spacing between the wires is greater than the wire diameter. This type of electrostatic screen may be useful when ventilation is required for the screened component to get rid of heat.

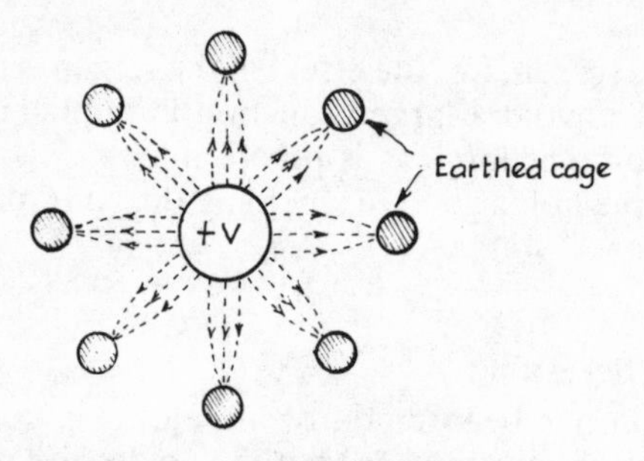

Fig. 9.22 Use of earthed cage (plan view).

An electrostatic screen is sometimes fitted between the primary and secondary windings of a mains transformer, see Fig. 9.23. This reduces capacitive coupling between primary and secondary which assists in reducing the transference of mains r.f. interference to the secondary circuits. It also reduces the possibility of the establishment of potentials

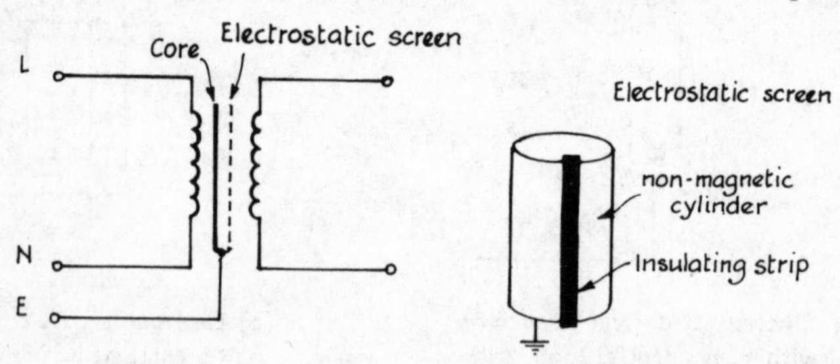

Fig. 9.23 Use of electrostatic screen in transformer.

referenced to 'live' or 'neutral' in the secondary which is important in a mains isolation transformer.

If the primary winding were to be enclosed by an earthed copper cylinder before the secondary were wound on, it would act like a short-circuited turn and prevent normal magnetic coupling between primary and secondary. To allow magnetic coupling while providing electrostatic shielding, all that is necessary is to place a strip of insulating material between overlapping edges of an earthed non-magnetic metal cylinder. It should be noted that an earthed copper screening can will act as both a magnetic and an electrostatic shield at r.f.

QUESTIONS ON CHAPTER NINE

(1) A transformer with a step-down ratio of 4:1 has 25V applied to its primary. The secondary voltage will be:
(a) 100V
(b) 6·25V
(c) 1·5625V
(d) 2·5V.

(2) A transformer with a step-up ratio of 1:8 has 60V in the secondary. The primary voltage is:
(a) 7·5V
(b) 4V
(c) 480V
(d) 320V.

(3) The secondary current of a transformer using a 10:1 step-down ratio is 100mA. The primary current is:
(a) 1mA
(b) 10mA
(c) 1A
(d) 100A.

(4) If the secondary load of the transformer in question 3 is 100Ω, the primary voltage will be:
(a) 1V
(b) 2·5V
(c) 10V
(d) 100V.

(5) The primary current of a transformer with a step-up ratio of 1:7 is 1·4A. The secondary current will be:
(a) 9·8A
(b) 0·707A
(c) 0·2A
(d) 8·4A.

(6) A transformer with a 9:1 step-down ratio has a secondary load of 5Ω. The resistance seen in the primary will be:
(a) 0·55Ω
(b) 405Ω
(c) 45Ω
(d) 5Ω.

(7) An amplifier with an output resistance of 245Ω is to be matched to a 5Ω load. The turns ratio of the transformer used will be:
(a) 1:49
(b) 7:1
(c) 49:1
(d) 1225:1.

(8) The core of a transformer is laminated to:
(a) Reduce magnetic coupling
(b) Increase the primary inductance
(c) Reduce interference
(d) Reduce eddy current loss.

(9) To provide magnetic shielding at low frequencies it is essential to use:
(a) A screening box of high resistance
(b) An earthed screening box
(c) A screening box of high permeability
(d) A copper screening box.

(10) A suitable material for magnetic screening at high frequency would be:
(a) Iron
(b) Ferroxcube
(c) Copper
(d) Silicon.

(11) To provide electrostatic screening it is essential to use:
(a) An earthed screening box
(b) An iron screening box
(c) A low-permeability screening box
(d) A cage.

(Answers on page 180)

CHAPTER TEN

MEASURING INSTRUMENTS

THE MEASURING INSTRUMENTS used in electronic servicing should be of a standard to detect reliably the actual voltage, current, resistance or waveform levels in equipment (within stated circuit tolerances), to enable the servicing technician to make decisions on the correct/incorrect operation of the various circuits.

TYPES OF INSTRUMENT

(a) Analogue

Instruments which have a pointer that is deflected over a scale represent the quantity being measured in analogue form since the magnitude of the deflection, which moves smoothly over the scale, represents the magnitude of the quantity.

The moving-coil instrument is an example of this type, see Fig. 10.1 and its operation was described in Volume 1 of this series. In a moving-coil

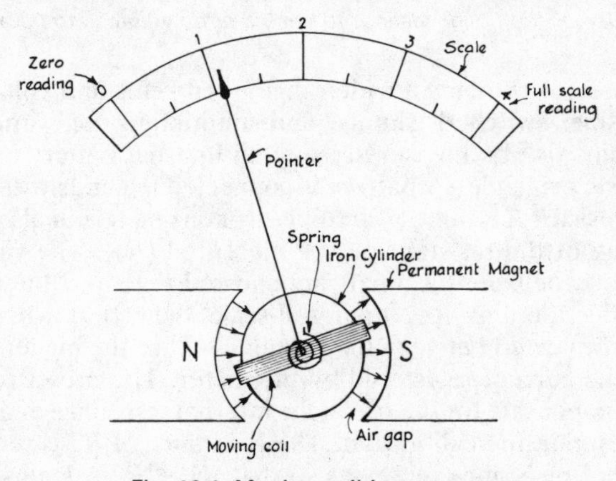

Fig. 10.1 Moving-coil instrument.

instrument the movement of the pointer is directly proportional to the **mean value** of the current flowing in the moving coil and the scale of the instrument is a linear one. The accuracy of measurement with this type of analogue display depends not only on the design accuracy (e.g. 1% from one tenth of full-scale to full-scale on d.c.) but also on the estimation of the pointer position relative to the scale. To reduce reading errors, a knife-edge pointer may be used and the scale is fitted with a mirror. By aligning the pointer with its reflection in the mirror, the scale is always viewed at right angles.

The moving-coil meter is a very sensitive instrument and may require only 20μA or 2mA for full-scale deflection. In order to read larger currents, a portion of the total current is passed through a resistor connected in parallel with the instrument and is called a **shunt**, see Fig. 10.2.

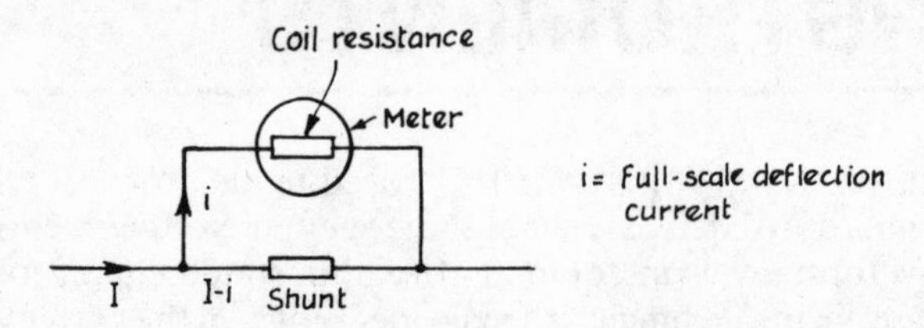

Fig. 10.2 Use of shunt resistor to enable larger currents to be measured.

The basic moving-coil instrument requires only a very small voltage applied to it (millivolts) to give full-scale deflection. To enable larger voltages to be measured a resistor is connected in series with the instrument called a **multiplier**, see Fig. 10.3

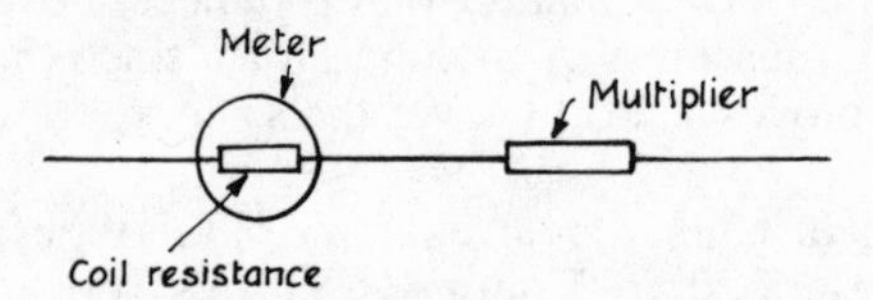

Fig. 10.3 Use of multiplier resistor to enable large voltages to be measured.

In 'universal' instruments a wide range of currents and voltages may be measured using switched shunts and multipliers of various values. Resistance may also be measured using an internal battery and Fig. 10.4 shows the basic principle. A battery is connected in series with the current meter and a resistor $R1$, but the circuit is broken and terminals provided for the connection of the resistance to be measured (R_x). The value of $R1$ is chosen so that when the terminals are shorted together, the meter reads full-scale deflection but less than full-scale deflection when an actual resistance is connected between the terminals. Thus the higher the value of R_x, the less the current registered by the meter. The preset resistor $R2$ is included to compensate for the rise in the internal resistance of the battery as it ages, to maintain the calibration. The resistance of $R2$ is reduced as the battery internal resistance increases so that $R1 + R2$ remains the same.

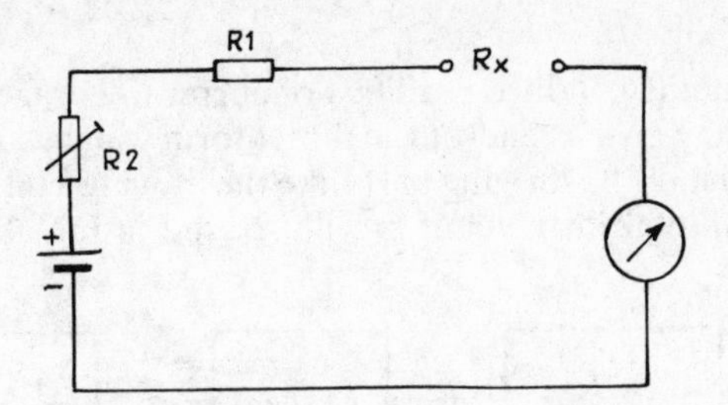

Fig. 10.4 Principle of ohmmeter.

The disadvantage of the moving-coil instrument is that it will operate directly on d.c. only since it is a mean or average reading instrument. However, by using a rectifier unit to convert a.c. into d.c. the instrument may be used to measure also alternating voltages and currents, see Fig. 10.5. Although the meter reads the mean value of the rectified a.c. it is normally calibrated in r.m.s. values. The r.m.s. calibration holds good only when measuring sinewaves.

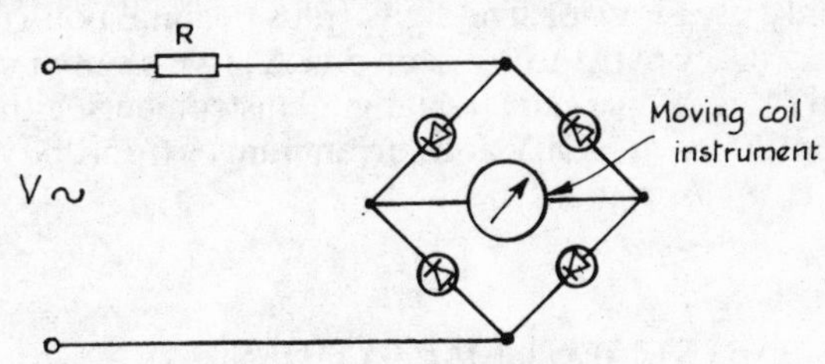

Fig. 10.5 Measuring a.c. with the moving-coil instrument.

A moving-coil instrument is used also in an 'electronic voltmeter' where the voltage to be measured is applied to an electronic amplifier with a high input impedance, see Fig. 10.6. D.C. or a.c. may be measured with a frequency range extending up to 1MHz or higher.

The ordinary c.r.o. discussed in Chapter 7 is another example of an analogue instrument used commonly in servicing work. Here the vertical deflection of an electron beam gives a continuous indication of the value of voltage or current to be measured. The c.r.o. is particularly useful for assessing waveform shape, peak or peak-to-peak values, frequency, periodic time, pulse period, phase difference and d.c. component of a.c. waveforms.

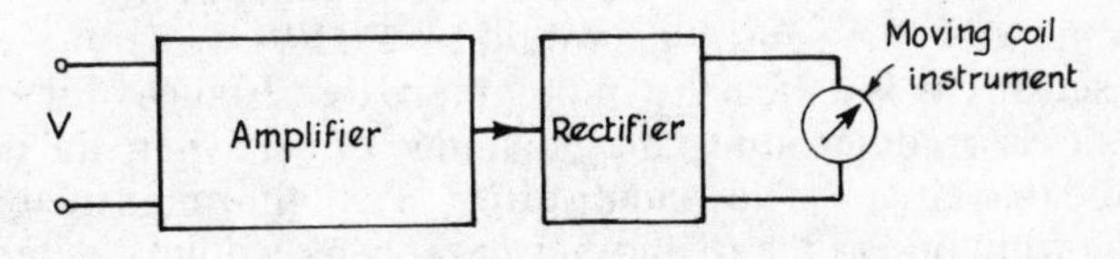

Fig. 10.6 Essentials of electronic voltmeter.

(b) Digital

Digital instruments produce readings of current, voltage or resistance in discrete steps. They give a read-out in direct form which is free from human reading error and have no moving parts like the moving-coil instrument. The basic ideas of a digital instrument are illustrated in Fig. 10.7.

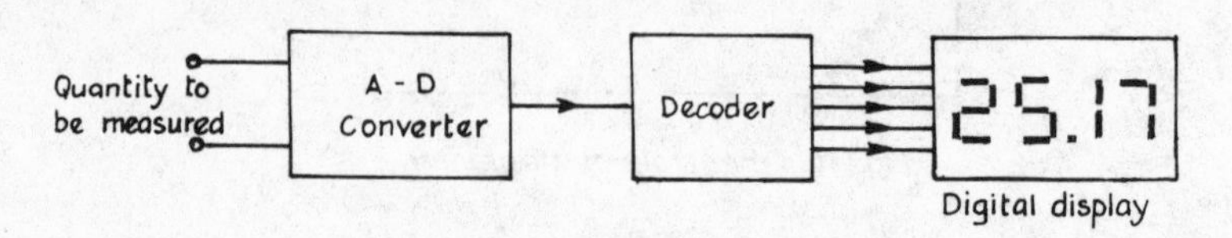

Fig. 10.7 Essentials of digital instrument.

The analogue quantity to be measured is converted into digital form, i.e. binary pulses, by an A–D converter. The pulse output of the A–D converter which is proportional to the analogue input is then decoded to operate the digital display consisting of seven segment devices (l.e.d. or liquid crystal). The display commonly uses four or five digits (plus decimal point) with an instrument reading accuracy of 0·1 to 0·8% on d.c. A large number of ranges are usually provided (e.g. 28) for universal digital instruments with voltage ranges extending from 200mV to 750V and current ranges from 20μA to 10A for both a.c. and d.c. measurements.

INSTRUMENT LIMITATIONS

All measuring instruments have certain limitations which should be considered in choosing an instrument for a particular application. The limitations include:

(a) Sensistivity; (b) Frequency range; (c) Max/min ranges; (d) A.C./D.C. current/voltage.

Sensitivity

The sensitivity of an instrument is given by $1/I_{fsd}$ ohms per volt. For example a meter reading 1V full-scale deflection which requires a current of 50μA must have an internal resistance of $1/(50 \times 10^{-6}) = 20\text{k}\Omega$. Thus the meter sensitivity is 20kΩ/V. A meter requiring a current of 1mA to read 1V full-scale deflection must have an internal resistance of $1/(1 \times 10^{-3}) = 1\text{k}\Omega$. Thus in this case the sensitivity would be 1kΩ/V.

Thus sensitivity is a measure of the internal resistance of the meter and is important as it determines the 'loading' effect when an instrument is connected into circuit. The loading effect of a multi-range meter alters when switching from one range to another but it can readily be determined if the basic sensitivity of the instrument is given, see Fig. 10.8.

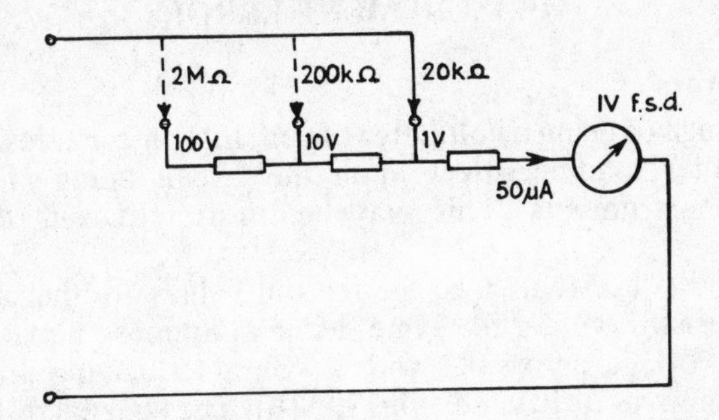

Fig. 10.8 Effect of voltage range on internal resistance of instrument.

In the figure it will be seen that if 50μA is required to give a full-scale deflection of 1V then the sensitivity of the instrument is 20kΩ per volt. Therefore on the 1V range the instrument has an internal resistance of 20kΩ. Thus on the 10V range it will be ten times 20kΩ (= 200kΩ) and on the 100V range one hundred times 20kΩ (= 2MΩ). Therefore the loading effect decreases as the range of the instrument increases and the most accurate measurement will be obtained on the highest range, in spite of the observed reading then being only a small fraction of f.s.d., a condition normally to be avoided.

Frequency range

Instruments will operate with their specified accuracy over only a particular frequency range, due to the effects of stray capacitance and inductance of components or bandwidth limitations of amplifiers used in the instrument. If attempts are made to measure signals with frequencies outside the specified range, low readings will generally be obtained but with occasional high reading at particular frequencies. Some typical frequency ranges for servicing-type instruments are given in Table 10.1.

Instrument	Input Resistance	Frequency Range
Analogue Universal Instrument	20kΩ/V	0–20kHz
Analogue Electronic Voltmeter	1MΩ/V	0–100kHz
Digital Universal Instrument	10MΩ on d.c. and a.c. ranges	0–50kHz
Analogue C.R.O.	2MΩ	0–15MHz

Table 10.1 Some instrument parameters

MEASUREMENT ERRORS

Voltmeter Errors

The advantage of using a voltmeter of very high internal resistance is that it has a smaller loading effect upon the circuit being measured than less-sensitive instruments. This may be illustrated using the circuit of Fig. 10.9(a).

It is easy to deduce that since the resistor values are the same, the p.d. across each resistor will be the same, i.e. 6V. Suppose now that we try to measure the voltage across *R*2 with a voltmeter having a sensitivity of 10kΩ/V and an f.s.d. of 10V, see Fig. 10.9(b). The internal resistance of the meter will be $10 \times 10\text{k}\Omega = 100\text{k}\Omega$ and since this resistance is in parallel with *R*2, the effective resistance of the combination will be 50kΩ. Thus, due to its loading effect, the voltmeter will read only 4V, an error of 33⅓%. It makes no difference if the voltmeter that is used is the most accurate instrument ever manufactured; its low internal resistance produces a large error in measurement when it is connected into circuit.

If a voltmeter having a sensitivity of 500kΩ/V and a f.s.d. of 10V is used, see Fig. 10.9(d), to measure the voltage across *R*2, the effective resistance of

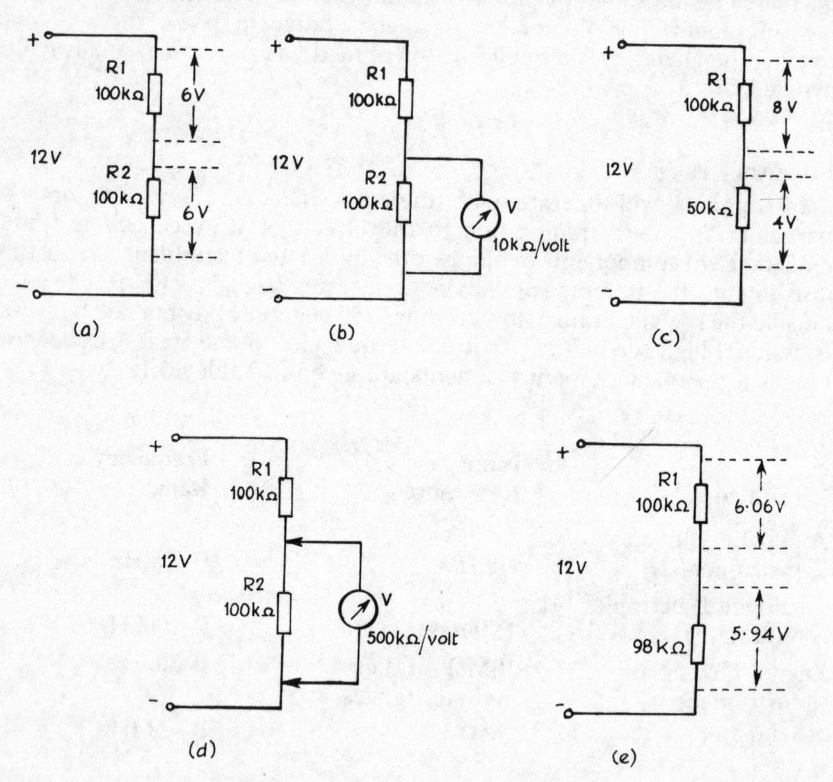

Fig. 10.9 Voltmeter error due to loading effect.

the combination will be 98kΩ. Thus the p.d. across each resistor will be as in Fig. 10.9(e) giving across *R*2 a voltage of 5·94V, an error of only 1%. In order to minimise errors of this type the internal resistance of the voltmeter used should be at least ten times the resistance (or impedance) of the circuit being measured.

Ammeter Errors

Although the resistance of ammeters is comparatively low, the introduction of an ammeter into a circuit can sometimes change the circuit resistance by an amount large enough to produce significant measurement error. An instrument manufacturer will specify either the resistance of the ammeter or the p.d. across the terminals when reading f.s.d. For example, the p.d. across the terminals of a typical analogue universal instrument is 0·5V on all d.c. current ranges at f.s.d. Thus, when on the 1mA range the instrument resistance is $0{\cdot}5/(1 \times 10^{-3}) = 500\Omega$; on the 10mA range it is $0{\cdot}5/(10 \times 10^{-3}) = 50\Omega$; and on the 100mA range it is $0{\cdot}5/(100 \times 10^{-3}) = 5\Omega$ etc.

Consider now that such an instrument is used to measure the current flowing in the circuit of Fig. 10.10(a). It may be calculated from the values given that the actual current flowing will be 1mA. If now an ammeter set on its lmA range is introduced into the circuit as in Fig. 10.10(b) it is equivalent to adding an extra 500Ω in series with the 2.5kΩ resistor, see Fig. 10.10(c). The measured current flowing will be 0·83mA, an error of approximately 17%. The error may be considerably reduced if the current is measured with the instrument set on its 10mA range. The resistance of the instrument will be reduced to 50Ω and the measured value will be 0·98mA producing an error of only 2%, see Fig. 10.10(d).

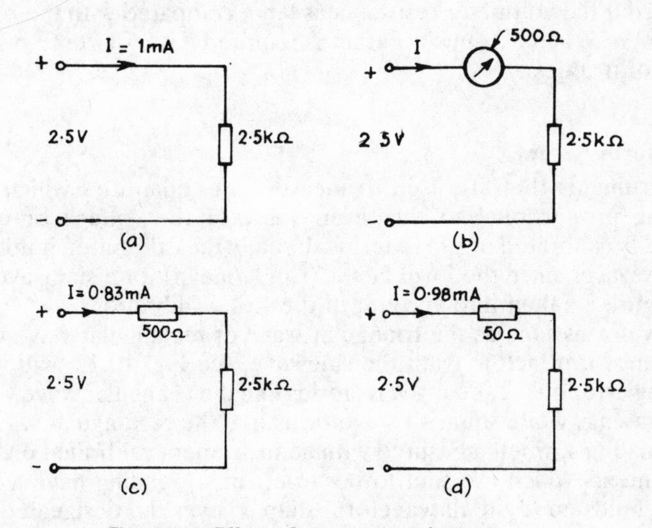

Fig. 10.10 Effect of ammeter resistance.

Clearly, the lower the ammeter resistance in relation to the circuit resistance the less will be the measurement error. For reasonable results the ammeter resistance should be no more than one tenth of the circuit resistance.

Voltage and Current Errors

Figure. 10.11 shows two ways of connecting a voltmeter and an ammeter to measure simultaneously the forward voltage drop and the current in a p-n diode. In (a) the ammeter will measure the true current in a diode but the voltmeter will indicate the voltage drop across the diode plus the voltage drop across the ammeter. Whereas, in (b) the voltmeter will indicate the true voltage drop across the diode but the ammeter will read the diode current plus the voltmeter current. For both cases a measurement error will occur

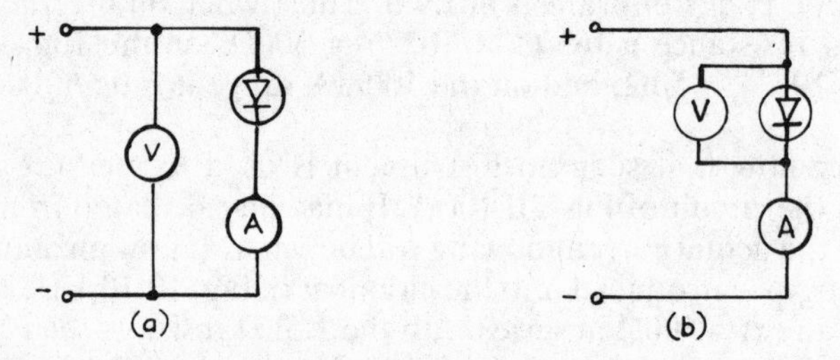

Fig. 10.11 Voltage and current errors.

but the percentage error will depend upon the instruments chosen. In (a) the voltage error will be reduced if an ammeter is chosen with a resistance that is small compared with the diode resistance, and in (b) the current error will be reduced if the voltmeter resistance is large compared with the resistance of the diode. When extreme accuracy is required it is important to be aware of these difficulties.

Waveform Errors

Instruments that are used to measure a.c. quantities which produce a reading proportional to the **mean value** of the voltage or current are normally calibrated in r.m.s. terms. Usually the calibration holds good only on sinewaves since the form factor (r.m.s./mean) for a sinewave is 1.1 and this figure is taken into account in the scale calibration.

Waveforms such as the triangular wave or rectangular wave which have different form factors from the sinewave, see Fig. 10.12, will introduce a reading error on a.c. Peaky waveforms like the triangular wave will produce low readings while squarish waveforms like the rectangular wave will give high readings, when measured with an instrument calibrated on sinewaves. Instruments which are said to be 'true r.m.s.' reading have a calibration which holds good on all waveform shapes, over the designated frequency range of the instrument.

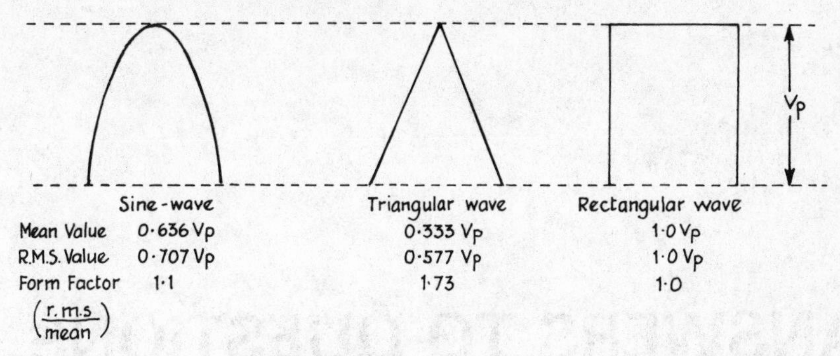

Fig. 10.12 Form factor of waveforms.

QUESTIONS ON CHAPTER TEN

(1) A moving-coil instrument produces a deflection proportional to:
(a) the peak value
(b) the mean value
(c) the r.m.s. value
(d) the square root of the peak value.

(2) The scale of the moving-coil instrument on a.c. is calibrated in:
(a) r.m.s. values
(b) peak values
(c) mean values
(d) average values.

(3) An instrument has a sensitivity of 50kΩ per volt. The internal resistance on the 10V range will be:
(a) 50kΩ
(b) 5kΩ
(c) 500kΩ
(d) 5MΩ.

(4) To reduce measurement errors with a voltmeter, the internal resistance of the instrument should be:
(a) equal to the circuit resistance
(b) one tenth of the circuit resistance
(c) one half of the circuit resistance
(d) ten times greater than the circuit resistance.

(5) One advantage of a digital voltmeter over an analogue one is that:
(a) it has a greater frequency range
(b) reading errors are reduced
(c) range accuracy is always better
(d) reads true r.m.s. on a.c.

(Answers on page 180)

ANSWERS TO QUESTIONS

Chapter 1

No. 1 (d)
2 (b)
3 (b)
4 (b)
5 (a)
6 (b)
7 (a)
8 (c)

Chapter 2

No. 1 (a)
2 (d)
3 (b)
4 (b)
5 (d)
6 (c)
7 (d)
8 (c)
9 (a)
10 (a)

Chapter 3

No. 1 (b)
2 (d)
3 (b)
4 (a)
5 (d)
6 (c)
7 (b)
8 (c)
9 (b)
10 (d)
11 (a)
12 (d)

Chapter 4

No. 1 (a)
2 (d)
3 (d)
4 (c)
5 (b)
6 (b)
7 (c)

Chapter 5

No. 1 (d)
2 (c)
3 (c)
4 (a)
5 (d)
6 (a)
7 (b)
8 (a)
9 (b)
10 (b)

Chapter 6

No. 1 (d)
2 (a)
3 (a)
4 (b)
5 (b)
6 (b)
7 (b)

Chapter 7

No. 1 (d)
2 (c)
3 (c)
4 (b)
5 (b)
6 (a)
7 (a)

Chapter 8

No. 1 (d)
2 (c)
3 (a)
4 (a)
5 (d)
6 (b)
7 (c)
8 (d)
9 (b)
10 (b)
11 (d)
12 (b)

Chapter 9

No. 1 (b)
2 (a)
3 (b)
4 (d)
5 (c)

No. 6 (b)
7 (b)
8 (d)
9 (c)
10 (c)

No. 11 (a)

Chapter 10

No. 1 (b)
2 (a)
3 (c)
4 (d)
5 (b)

INDEX